The World of Science Fiction Films

by

Satyen Mohapatra

The World of Science Fiction Films
by Satyen Mohapatra

ISBN: 978-93-59956-88-6

Published by

DOUBLE 9 BOOKS

2/13-B, Ansari Road
Daryaganj, New Delhi – 110002
info@double9books.com
www.double9books.com
Tel. 011-40042856

ABOUT THE AUTHOR

Satyen Mohapatra is a highly experience journalist who has worked for over 30 years for the Hindustan Times which is one of the leading English daily newspapers in India. He is accredited as Senior Journalist with the Press Information Bureau of the Government of India. While he has long been doing off the beaten track and human interest stories in various subjects, over recent years he has specialised as a science and features writer. He has also taught college students and in a business management institute. Mohapatra has won several of awards for his news stories and articles. He has authored an illustrated reference work Delhi: A City of Museums. Mohapatra's exposure to science and to scientists through his work as a journalist, and his interest in science fiction films led him to write a book on the subject. In his book Satyen Mohapatra has covered a broad canvas. He begins by defining science fiction, differentiates it from fantasy and discusses how it can help to ponder over the deep questions and issues that face humankind. After the general discussion he provides a history of science fiction films and describes the life and work of ten leading western directors of the films, including famous names like Steven Spielberg, James Cameron and Christopher Nolan.

CONTENTS

Foreword

Ajit Kembhavi

Science fiction is imaginative literature based on known scientific laws, and so are science fiction films. Science fiction is different from fantasy, which is limited only by the imagination of the creator. The worlds of fantasy adhere to their own laws, a profound example of such a world being the Middle-earth of J.R.R. Tolkien. Science fiction of course needs to stretch known scientific laws to their limits, and even to extrapolate them for the work to be interesting, so the boundary between fact and fiction becomes dented and blurred. But unlike fantasy, the science fiction of one technological epoch can become the reality of another, of which there are many examples over the twentieth and the present century.

My first exposure to science fiction was through the works of Jules Verne and H. G. Wells.

Some of the ideas introduced there like advanced submarines and travel to the Moon have already been realized, a journey to the centre of the Earth will perhaps forever remain impossible, while time travel remains just a hope for the distant future. The first modern science fiction story I read was Nightfall by Isaac Asimov. That wonderful story, first published in 1941, involves intelligent beings on a planet outside our Solar system, orbiting a system of two stars. Such a system is scientifically possible, but in 1941 no planet outside the Solar system was known. The first extrasolar planet was discovered by astronomers in 1995 and more than 5000 such planets have so far been found. These account for only a very tiny fraction of the billions of planets now believed to exist in our galaxy. A fraction of these have Earth like conditions, so life like ours, or more exotic life as in the film Avatar is now wholly plausible on a staggeringly large number of planets. Wars between the Earth and other worlds, which are the lifeline of so many stories and films, may very well take place someday. But for those to happen we will have to first solve the problem of getting from one world to another.

We know from Albert Einstein's work that nothing can travel faster than light, and even light can take thousands of years even to travel to moderately distant parts of our galaxy. To reach the distant worlds in our galaxy in any reasonable time we will need to breach the limit of the speed of light, for which we will have to discover completely new physical principles which go beyond what we presently know. But Einstein's theory of gravity may provide another way out. In the curved space-time introduced by Einstein there can be wormholes or bridges which provide shortcuts to distant parts of the universe, like tunnels which quickly take us across a mountain, avoiding a long circuitous route around it.

Presently such possibilities are in the twilight zone between proven science and its more fanciful extensions, and are therefore just right for the creators of science fiction to base their stories on. Getting to the distant worlds and back, particularly to the more extreme ones, can have unintended consequences, like the young father meeting his aged daughter in the film Interstellar.

Astronomy and physics are not the only sciences which inspire fiction, a case in the point being the film Contagion of 2011, which was based on the rapid spread of a lethal virus, the story of which so closely resembles the pandemic we are currently living through. All the sciences have their own tales waiting to be told, and science fiction films provide a very powerful and entertaining medium for the narration. Films go far beyond the written word through the visualisation they provide, which can be particularly important for the abstract and unfamiliar themes that science fiction is often based on. I can quite easily imagine ungainly war machines trundling through an alien landscape; but I certainly need the help of colourful visual depictions, with a dramatic background score, to understand what it may be like to fall into a black hole.

Satyen Mohapatra is a highly experience journalist who has worked for over 30 years for the Hindustan Times which is one of the leading English daily newspapers in India. He is accredited as Senior Journalist with the Press Information Bureau of the Government of India. While he has long been doing off the beaten track and human interest stories in various subjects, over recent years he has specialised as a science and features writer. The subjects he has covered include the BT Brinjal Controversy, India's Moon Mission, Antarctic Expeditions and Science Education Policy. Mohapatra has interviewed prominent persons, covered the Rajya Sabha for fifteen years, has worked in senior management positions, as an editorial consultant, in news management, and in scripting and presenting radio programmes. He has also taught college students and in a business management institute.

Mohapatra has won several of awards for his news stories and articles. He has authored an illustrated reference work Delhi: A City of Museums. Mohapatra's exposure to science and to scientists through his work as a journalist, and his interest in science fiction films led him to write a book on the subject.

In his book Satyen Mohapatra has covered a broad canvas. He begins by defining science fiction, differentiates it from fantasy and discusses how it can help to ponder over the deep questions and issues that face humankind. After the general discussion he provides a history of science fiction films and describes the life and work of ten leading western directors of the films, including famous names like Steven Spielberg, James Cameron and Christopher Nolan. He then devotes a chapter each to descriptions of some famous science fiction films from other countries and from India. The latter part of the book deals with techniques used before the computer age for producing the rich imagery needed for the films, and how the extravagant effects which are now commonplace are produced with modern technology. The book ends with a discussion on how many of the ideas which first appeared as science fiction have later become a part of reality.

I believe such a book is very timely. It provides a perspective on an important genre of films, the appreciation for which is not yet mainstream. Science fiction films are mainly viewed as thrillers and the scientific nuances and asides that they may contain are often lost. The book helps to develop some insights into these subtleties which otherwise would go unnoticed. The book also helps to bind together seemingly different films, and to trace the evolution of ideas and techniques over the years, and from one director to another. I am sure the book will provide much instruction and pleasure to those who come across it, and will help them watch science fiction films with greater appreciation in the future.

Professor Ajit Kembhavi, Professor Emeritus and former Raja Ramanna Fellow and Founder Director at the world class Inter-University Centre for Astronomy and Astrophysics,Pune. He was Vice-President of the International Astronomical Union and former President, Astronomical Society of India, Chair of the International Virtual Observatory Alliance, Chair of the Scientific Councilof the Astronomical Data Centre at Strasbourg and Chair of the Council of the Indian Institute of Astrophysics and was on the Council of the Indian Institute of Science. He is a Fellow of the Indian Academy of Sciences and the National Academy of Sciences, India.

PREFACE

Satyen Mohapatra

As a twelve-year-old teenager in the mid-1960s, I was fascinated with science fiction.

The first exciting and amazing Sci-Fi story I read in the new and popular Hindi children's magazine of that time, "Nandan", left me completely awestruck. It was about a submarine with the crew shrunk to microscopic size placed in a syringe and injected into the bloodstream of a man.

Wow, what a flight of imagination. Miniaturized humans in a submarine heaving and surging on the high waves of swirling blood fearful of being gobbled by huge white corpuscles in the bloodstream. It was a high adventure.

I had not read that kind of stuff before. I was permanently hooked to Science Fiction.

The story was based on the storyline of "Fantastic Voyage" the 1966 American science fiction adventure film directed by Richard Fleischer based on a story by Otto Klement and Jerome Bixby.

As a highly imaginative kid in school and a science student, I continued my love for science fiction. No wonder after doing my Master in English Literature and taking up a journalist's job with the national daily The Hindustan Times, as its Special Correspondent, I was soon handling the Science beat.

It gave me a lifetime opportunity to interview innumerable Nobel laureates in science, interact with scientists, and do in-depth studies about different disciplines of science and scientific institutions. My interest was in many areas including Space, Biotechnology, Oceanography, Particle Physics, Atomic Energy, Zoology, Computer Sciences, Earth Sciences, Medical and Health Sciences, and Chemical Sciences.

The toughest part was to convey complex scientific information in simple and jargon-free language to the readers. A task I successfully performed for years on end and therefore while working on this book, I not only enjoyed the science part of it, but it was also highly enlightening to see how the creative mind of writers came up with interesting story ideas and fictional plots and characters from the major scientific and technological developments and breakthroughs.

For a Science Correspondent to take up the subject of science fiction films for writing a book did not require much of an inspiration and came naturally once I started looking for a subject to write about.

I always loved watching films like everyone else including science fiction films . One of the greatest and perhaps the first thoroughly enjoyable sci fi movie I watched in my childhood was "Planet of the Apes".

It was a shocking and thrilling experience particularly to see Charlton Heston of Ben-Hur fame cowering and fleeing from gun-toting huge apes riding on horses. The complete reversal of the Man-Animal relationship shown in the film was scary and remembering those vivid scenes even after decades give me goosebumps.

Another impactful film that I remember from those days is the sci-fi classic 2001 Space Odyssey. Made on a huge canvas with technological finesse and strong philosophical and spiritual underpinning. It was awesome.

An extremely moving film even after decades its unique position has not been challenged and being philosophically inclined, I have no hesitation in calling it the greatest sci-fi film. With a colossal sweep director, Stanley Kubrick's thought-provoking movie takes us from Stone Age to Space Age and beyond giving a stunning glimpse of human destiny – a visual masterpiece.

Deeply interested in philosophy I found during my research on sci-fi films that almost all the great science fiction films were basically dealing with perennial philosophical questions pertaining to human nature and human destiny.

Who are we? What does the future hold for us? What is the nature of Reality? What is Mind? Are Humans capable of dealing with Intelligent Machines? What is Consciousness?

It was really very exciting and fun to work on other interesting areas and questions raised in sci-fi films like - Is time travel possible? If man could be miniaturized? If we could travel to other planets? What would aliens of another planet be like? Would aliens be friends or enemies? Are there other physical dimensions? If dinosaurs were alive today? Can man become a superhero? What would society on earth be like in the future? If robots and AI take over the earth? Can Man have an intimate relationship with a robot?

Innumerable sci-fi films have been made taking up these questions in one way or another. I have written a full chapter on Philosophy and Sc-Fi films.

It was also very educative to learn about the different technologies like Computer Generated Imagery, Schufftan Process, Prosthetics, and Motion Capture technology used to create those out-of-the-world scenarios and dramatic moments.

What motivated me was also the fact that there were not many books available on science fiction films with a chapter on Indian sci-fi films, though books on science fiction itself could fill libraries. I was keen to fill the gap.

If there is a profusion of science fiction books, I do not understand why books on science fiction films can be counted on my fingertips, particularly when large numbers of science fiction films are produced annually. Not only that most of the international blockbusters nowadays turn out to be sci-fi.

The growing popularity of sci-fi films mostly Hollywood films among youngsters in India is so much so that most of the Hollywood sci-fi blockbusters are being dubbed not only into Hindi but also into other regional languages.

The youth are finding the extraordinary challenges, the futuristic scenarios of science fiction films more appealing than the humdrum family drama/romantic comedies or crime movies, as the growth of CGI (Computer Generated Imagery) has greatly enhanced the level of special effects and completely transformed the viewing experience. Earlier science fiction films were stagey and amateurish compared to the films today.

As science fiction is a vast field, I could not do justice to it if I did not limit its scope. Therefore, not getting into a detailed discussion on the semantics of science fiction, I tried to keep myself within a strict definition of science fiction.

The definition which I have tried to follow is that science fiction can be any speculative and imaginative work that is based on hardcore science. It cannot degenerate into magic and fantasy and then call itself science fiction. However high the flights of creative fancy may take you; you must have clear links tying you down to the scientific principles and developments which gave rise to those exotic ideas.

A large number of well-known and very popular films did not make the mark as true-blue science fiction films according to me because when one carefully assessed them one found the films to be pure fantasy or magic.

Even though many of them are being distributed as science fiction films and viewed as such by viewers, I had to perforce keep them out.

These kinds of problems crop up because the definition of science fiction itself is not static but dynamic. Some practitioners of science fiction have made its definition so broad that magic, horror, fantasy, and all kinds of creative genres have been palmed off as science fiction.

This explanation I hope will satisfy those readers who may not find many renowned science fiction films without a mention in my book.

The films produced in the sci-fi genre worldwide today are voluminous. I have just tried to talk about a few known films to highlight the different interesting aspects of sci-fi films.

My idea is just to give a taste of science fiction films to the reader. To hold the reader's hand and take him or her through the fascinating world of science fiction films.

What makes a film a science fiction film? What is the history of sci fi films? What are the kind of subjects are dealt with in sci-fi films ? The technical marvels behind making a sci-fi film? Who are some of the renowned directors of sci-fi films?

One interesting aspect of science fiction is that uncannily many of science fiction works have been able to predict the scientific and technological developments of the future even hundred or more years away almost accurately. I have written a short chapter on predictions in science fiction that have come true.

Indian science fiction films even though we toyed with the sci-fi genre very early in the early 1960s have always been few and far between. Even today when youngsters are lapping up science fiction movies all over the country somehow Indian film industry finds itself constrained primarily due to finances because the initial investment of a sci-fi film due to CGI and special effects is huge.

Another factor that weighs heavily on the heads of producers of sci-fi films is that large masses of Indian film viewer is still not very comfortable with "science" as a subject itself. No wonder producers see to it that customary song and dance sequences, comic interludes, and car chases with a villain (part of the typical formula for a successful Indian blockbuster) are introduced even in a sci-fi movie to ensure its success in the Indian scenario.

Both in Hindi films and in regional films it was really disappointing to find out that Indian cinema has actually very few serious science fiction movies which will strictly fall in the definition of sci-fi films which one has fixed for this book. Most films touted as science fiction have fantasy and magic as the drivers of their action rather than science which sadly at times is given a perfunctory treatment.

Science fiction and science fiction films dealing with cutting-edge science and technology and futuristic concepts are a genre primarily for students and young adults. The multi-billion-dollar Hollywood movie industry globally targets this young group with its series of sci-fi, superhero, and fantasy movies.

Now let me let you on to a secret, one day a young couple told me, "You, can't write a book on science fiction films" obviously they saw my grey hair and that was the day I put pen to paper.

CHAPTER 1
WHAT IS SCIENCE FICTION

Human beings are imaginative. It is through their imagination that they transcend their physical limitations of both time and place, to experience even though vicariously much more than what is possible in their lifetime.

Stories, tales, heard from mothers or grandmothers based on fantasy and myths spark the fire of imagination and creativity in the child.

As the child hears of the brave and courageous prince, the powerful and villainous giant, the flying horse, the beautiful and intelligent princess, the perilous voyage across the seas, the clash of the forces of good against evil, and the eventual success of the forces of good, the child with his innate vivid imagination and sense of wonder lives through the whole experience. This provides him with both joy and satisfaction along with a strong moral foundation, a basic framework of values that helps him to tackle the world outside once he grows up.

The magical world that the child builds up in his mind may have flying carpets, fire-spitting dragons, towering vicious giants, talking animals, and beautiful fairies and all fantasy and not related to real life.. The logic may be completely lopsided from an adult's point of view. But it is through these imaginative stories that the children make sense of the world they have come into, their critical thinking skills and problem-solving abilities are strengthened, and they know of good and evil, joy and sadness, wisdom, and stupidity.

Whether it is stories from the Panchatantra, Aesop's fables, or Arabian Nights, once the outer trappings of these fables are removed one would find they all deal with the human condition, human emotions, human ambitions, human problems, and solutions.

Automatically the child, though engrossed in the outer peripheral story of adventure and fantasy understands and learns the embedded age-old wisdom of goodness, courage and hope.

Myths, folk tales, fantasy and adventure stories both in the oral and written tradition play an important role in carrying forward civilizational values, traditions, knowledge from generation to generation.

Literature in the shape of novels, stories, poetry, and drama, plays an extremely important role in society as the repository of its creative expression.

Director and Author of the Upper School Literature Curriculum at Memoria Press, David. M. Wright, writing on "Why Read Literature", says, "Reading great literature exercises the imagination. We enjoy stories; it is a pleasure to meet characters and to live in their world, to experience their joys and sorrows. In a practical sense, an active imagination helps us perceive truth, make value judgments, and deal with the complexities of life in creative ways. It even aids in our ability to use logic and to reason well."

"Reading literature enables us to see the world through the eyes of others. It trains the mind to be flexible, to comprehend other points of view—to set aside one's personal perspectives to see life through the eyes of someone who is of another age, class, or race. Reading literature nurtures and develops the power of sympathetic insight."

The most important aspect of literature, he points out, is "literature helps us to know ourselves—in short, to understand man. For the subject of literature is man. In its pages, we learn about our creative and moral faculties, our conscience, and most importantly, our soul. We see man at the height of his glory and the depth of his folly—with every heartrending thought, action, emotion, and belief in between. In other words, literature holds a mirror up to human nature, revealing its inner depths and complexities, its array of virtues and vices; and moreover, it holds a mirror up to a cultural age, illuminating its shape and ethos."

It is these lofty aims of literature that the truly great works of science fiction (which is now a part of mainstream literature) fulfill to a large extent.

A precise definition of science fiction may continue to elude us, as perhaps the definition of any other literary genre. However, we must make an honest attempt to find out broadly what science fiction means.

Even though science fiction writers and critics have slightly differing views on it emphasizing one aspect or another.

Instead of 'science fiction' for the first time when the word is supposed to have been used it was called "scientifiction" by Hugo Gernsback a Jew from Luxembourg who emigrated to US in 1904, when he was 20 years old. He became an American inventor, writer, editor and magazine publisher. He also played an important role in popularizing the genre of science fiction by starting science fiction fandom and organizing the Science Fiction League.

He brought out the first science fiction magazine called "Amazing stories". He also created the term "science fiction", though he preferred the term "scientifiction".

"By 'scientifiction', he said, "I mean the Jules Verne, H. G. Wells and Edgar Allan Poe type of story—a charming romance intermingled with scientific fact and prophetic vision... Not only do these amazing tales make tremendously interesting reading—they are always instructive. They supply knowledge... in a very palatable form... New adventures pictured for us in the scientifiction of today are not at all impossible of realization tomorrow... Many great science stories destined to be of historical interest are still to be written... Posterity will point to them as having blazed a new trail, not only in literature and fiction, but progress as well."

Even though calling it scientifiction, he had basically got it right as far as the essence is concerned of what science fiction means to the lay reader even today. His vision of science fiction over the years has been prophetic.

The fact is that it is the human protagonist of the science fiction story or novel whose fate we as readers are more concerned about even in the most hardcore science fiction story. This is brought out well by famous science fiction writer, Theodore Sturgeon who defines science fiction as "A science fiction story is a story built around human beings, with a human problem, and a human solution, which would not have happened at all without its scientific content."

Emphasizing the point that science and technology is still the key element, novelist and poet, Sir Kingsley Amis says science fiction is "that class of prose narrative treating of a situation that could not arise in the world we know, but which is hypothesized on the basis of some innovation in science or technology, or pseudo-science or pseudo-technology, whether human or extra-terrestrial in origin."

Science fiction writers do not just talk about the possible scientific and technological advances, but their creativity lies in the ability to predict its impact on us. As well known science fiction writer Frederik Pohl puts it "Someone once said that a good science-fiction story should be able to predict not the automobile but the traffic jam. We agree".

In a similar tone, great science fiction writer Isaac Asimov says," Science fiction can be defined as that branch of literature which deals with the reaction of human beings to changes in science and technology."

Actually, actor Christopher Evans (Captain America), gives a much better definition encompassing a wide array of science fiction writing.

He says, "Perhaps the crispest definition is that science fiction is literature of 'what if?' What if we could travel in time? What if we were living on other planets? What if we contacted alien races? And so on. The starting point is that the writer supposes things are different from how we know them to be."

Renowned science fiction writer Robert Heinlein says, "It is always hard to face up to a complex world, try to figure out what makes it tick, try to cope with it, survive and triumph over it. But this is precisely what science fiction strives toward … Through science fiction humanity can try experiments in imagination too critically dangerous to try in fact".

Science fiction is based on imagined, future scientific or technological advances which lead to major social or environmental changes like computers ruling the world, man evolving into some other creature, space or time travel and life on other planets.

But then science or scientific development must be the key mover in any science fiction story or novel and must affect the lives of the protagonist.

Any imaginative story or fantasy therefore does not become part of science fiction if science or scientific principle or scientific invention is not involved in the development of the story.

Growth of technology along with its impact on society and culture will continue to inspire science fiction writers to come out with a creative response by penning great literary works taking on the present-day themes as well as the universal and eternal big questions.

Science fiction uses elements of science and technology in some way or another. The technology may or may not exist but there should be little doubt left in the minds of the readers that it is achievable in specific circumstances in the future, like technology of time travel and space travel. Newer forms of technology which are yet unheard or unthought of may come up in science fiction as part of extrapolation or speculation of the writer.

A world in another part of the universe peopled with life form, which is not necessarily carbon based but on silicon, palladium, germanium, titanium. Life may also be based on chemicals like hydrogen fluoride, hydrogen sulphide, ammonia or even methane.

As scientists toy with the idea of the possibilities of life based on some other element than carbon in some other part of the universe, which cannot be ruled out theoretically, nothing can stop science fiction writers to speculate on what if there actually exists some form of life based on lets say methane or arsenic.

It is actually themes and issues on the cutting edge of science and technology where scientists are also just speculating and coming out with their possible theories to explain the intriguing natural phenomena that alert science fiction writers who really go ahead and as a part of their "thought experiment" create in the context of the above example the fictitious methane or silicon-based life forms to people their planet in the galaxy far far away.

A large body of science fiction has also always served as a warning to the possible fall out of the advancement in science.

Being a double-edged sword science can be beneficial as well as harmful. Looking at the propensity of egotistical Man science fiction writers often create a dystopian world either in a futuristic setting in this world or in far-flung worlds in some other part of the Universe demonstrating how ignorance, pride, greed, and hate and violence devastates the world.

Generally the philosophical and moral questions raised in science fiction works by the protagonists are similar to what we are dealing with today in the real world. The dystopian world is usually showcased as something that humanity must avoid at all costs.

Interestingly human history shows us that the human imagination is always two steps ahead of human development. Therefore, elements of science fiction are seen in the stories of ancient times which are part of our myths and epics even though modern day science had yet to develop. Those stories are dubbed as proto science fiction.

There are several different genres of sci fi like alien invasion, time travel, alternate future or alternate history, scientific accident or invention leading to transformation at the individual level or any kind of socio-political or environmental change.

Science fiction also deals with the entire Man-Machine interface particularly in view of growing automatons. The nature of humanity or the fascinating question of what makes us human can be dealt with in the best way by a comparison with robots and androids.

Similarly, Man-Alien relationship also comes under the ambit of science fiction. Even though existence of aliens is not yet established by science, the idea can neither be dismissed and it gives sufficient fodder for science fiction writers to come out with innumerable stories and novels dealing with alien encounter, alien civilizations, alien physiology, alien psychology , alien philosophy. How will Man react to an alien? What kind of relationship Man will establish with aliens – a conqueror or conquered?

Science fiction is a genre of fiction in which the stories often tell about science and technology of the future. Its texts are often set in the future, in space, in a different world, or in a different universe or dimension altogether.

The moot point however is science fiction too must be based on principles of science. It can be imaginative and fictitious but should not be completely unbelievable, because it then ventures into the fantasy genre.

It also must include the human element, explaining what effect new discoveries, happenings and scientific developments will have on humanity in the future. Though it can also be entirely about aliens in some other part of the Universe then the writer must be able to engross us sufficiently with the story of the protagonist, which we know will be possible only if he/she falls into somekind of human predicament with which we can connect in our mind.

Science fiction must contain the usual elements of the story/novel - a specific location, theme, character , plot, conflict and complications and final resolution.

Isaac Asimov in an excellent tribute to science fiction says, "Modern science fiction is the only form of literature that consistently considers the nature of the changes that face us, the possible consequences, and the possible solutions...That branch of literature which is concerned with the impact of scientific advance upon human beings."

According to well known science fiction writer Robert A. Heinlein science fiction is a realistic speculation about possible future events, based solidly on adequate knowledge of the real world, past and present, and on a thorough understanding of the nature and significance of the scientific method.

What the science fiction writers says is that even though their writings are speculative it is based on the real world and does not in any way negate the established facts and natural laws. Any flight of fancy would follow rigorous reason and possibility in the real world.

r.

Actually being speculative, it is free from time and space constraint, the authors can write about any time in the past or present or future and the event can be situated anywhere in space throughout the known universe and in any dimension.

Science fiction writing is about how life is impacted by technological change. Life can be shown in the alternative worlds but universal laws of Nature cannot be flouted even there.

Science fiction also speculates about changes that can take place in human beings themselves. Will we become cyborgs with technological development? Will we be communicating through telepathy in the future? Will humans evolve into something else completely different? Why will we change, how will we change? How will we cope with the physical, mental or genetic changes taking place at great speed which may completely transform our present being? Will our consciousness or our spirit remain the same or change too?

These are some of the questions in the arena of science fiction writers.

One of the greatest benefits of science fiction is that it allows us to go through disasters, alien encounters, out of the world scenarios, in our mind's eye. The whole question of "what if" one of the scenarios happened is answered for us without actually having to go through anything of the kind.

Science fiction because it is based on scientific method and logic (even though extremely far fetched at times) cannot be rejected outright for being nonsensical or just a fantasy.

Thus, science fiction actually provides us innumerable ideas at times as to how science can impact human beings themselves and their environment, which potentially can help us in our understanding and decision-making process in the real Universe.

Sophia Brueckner, MIT professor, has lamented the fact that researchers whose work deals with emerging technologies like genetics are often unfamiliar with science fiction.

"I feel reading their writing can be just as important as reading research papers."

Time travel stories open up the whole arena of human history, time travelers by their action make changes in a specific point in history which can lead to a completely different outcome (utopian or dystopian) of a known historical event.

For example, if the assassination of Archduke Franz Ferdinand of Austria-Hungary could have been avoided would the World War I taken place. In, The Man in the High Castle, award winning novel by sci fi writer Philip K. Dick comes out with alternate history where Germany and Japan have won the World War II.

Similarl, predictions are played out. With status quo continuing as it is and if environmental degradation continues at the present rate what would happen a thousand or two thousand years afterwards?

Science fiction many a times deals not with individual people and their adventures and experience but about the human species as a whole as they undergo change due to progress in the different fields of science.

Science fiction is also philosophical fiction that ponders on big questions, such as "What is it to be human?" and "What is our real purpose on this planet?" Changed perspectives – distance in time, space, culture, and so forth – grant the science fiction writer speculative opportunities to consider many urgent philosophical themes.

It is behind the mask of science fiction writing and ruse of depicting imaginative cultures in some other planet or dystopian an imaginary society where people suffer from misery and oppressive social control in the future that writers have expressed their dissidence against the present dictatorial regimes in their own countries, which they cannot point out openly. .

Interestingly many scientists or engineers were inspired by reading or watching science fiction films to take up their careers.

While there's no single accepted definition of science fiction, science fiction usually deals with worlds that differ from our own as the result of new scientific discoveries, new technologies, or different social systems.

In the early civilization whether it is of Egyptian, Chinese, Indian as men settled down near rivers leaving the role of hunter-gatherer to take up agriculture and cultivation, they also built a collection of epics, myths, and stories to be transmitted from generation to generation talking of gods, monsters, and fantastic voyages to far-off places.

All over the world these kinds of tales demonstrating the creative imagination of human beings to explain away the natural phenomena are to be found.

Usually in the early days of human civilization all kinds of natural phenomena were attributed to powerful gods and monsters. The links to early myths to religion was very close. We have mythical gods like rain god, sun god, sea god,and evil monsters of all kinds with dragons and giants.

In India we even have stories connected with asuras like monsters Rahu and Ketu temporarilygobbling up the moon and sun to explain away the phenomenon of lunar and solar eclipse.

If one looks at the ancient epics particularly the Ramayana and the Mahabharata they are full of stories which resemble science fiction.

Whether it is the Pushpaka Vimana - the flower bedecked chariot flying through the air controlled by thought, teleportation by rishis, weapons like brahmastra with destructive power like nuclear war heads, test tube babies, television. The Indian epics are replete with these shrouded in the mythical language, characters and symbols.

Indian rishis and scholars are supposed to have knowledge of atom, gravity, speed of light in the Vedic era as interpreted in several shlokas now coming to light.

The television technology seems to have been in use during the Mahabharata.

The entire Mahabharata war was seen by Sanjay as if on live television once he gets 'DivyaDristhi' (sounds familiar like DD of Doordarshan). Thus,Sanjay saw whatever was taking place in the field of Kurukshetra and gave a live commentary to King Dhritarashtra who was blind but wanted to know what was happening in the battlefield.

The way it has been described with Sanjay having the power to go to specific areas in the battlefield on Dhritarashta's urging to relate different sequences seems as if there were different cameras fixed at different places to provide audio visual feed or there were cameras equipped with drones with some kind of a satellite link up to King Dhritarashtra's palace.

Interestingly there seems to have been an instance of posthumous sperm retrieval technology also seen in Indian myths particularly Vishnupurana.

Nimi was the son of Ikshavaku who started a dynasty that had as descendants famouskings including Ramachandra, Trishanku, Harishchandra, Sagara, Bhagiratha.

Nimi desired to hold a yajña and requested Guru Vashiṣṭha to act as high priest at the yajña, as Vaṣhiṣṭha was busy with another yajna he asked Nimi to postpone it. However, Nimi went ahead with performing the yajna with Gautama as high priest.

When Vashishtha became free he came to Nimi to conduct his yajña but found that the yajna had already been held, angry at this Vashiṣṭha cursed Nimi that he should lose his body and become Videha.

This resulted in chaos in the state as Nimi did not have an heir and now lay dead.

King Nimi was preserved by being embalmed with fragrant oils and resins.

The dead body was placed in a churn in hope that the dead body of Nimi would re-assume human form. Sages succeeded in their effort and soil emerged from the churn. They created a son out of that soil from his dead body, who was called Janaka.

This case is known as a case of posthumous sperm retrieval which is just a speculation even today.

The early Greeks also had many myths in their epic stories which would go as protoscience fiction. In Jason's journey aboard the Argo he talks to a beam built into the Argo which could have been some kind of a robot,

In the second century work, True History by Lucian of Samosata, an Assyrian satirist, depicts encounters with alien life forms from the Sun and the Moon, and also interplanetary warfare for colonizing the Morning star, trees that look like women, and there is an incident of being whisked away to the Moon by a whirlwind.

Antonius Diogenes in 'Of the Wonderful Things Beyond Thule' has also recorded in the adventures of the Greek explorer Deinias who goes as far as the Moon.

In The Adventures of Bulukiya, from the Arabian Nights the hero travels across the cosmos, to worlds inhabited by talking snakes and trees. Abdullah the Fisherman and Abdullah the Merman stories describe an underwater society that practices primitive communism, and mechanical life forms are mentioned in The City of Brass, The Ebony Horse, and Third Qalandar's Tale.

It goes to the credit of Sir Thomas More in 1516 to introduce the word "Utopia" or ideal society, which was a concept adored by many science fiction writers as they projected alternative worlds or societies either good 'utopia' or bad 'dystopia' in their novels. .

Sir Thomas More published 'Utopia' describing a fictional ideal island society called Utopia where there is no private property, population is controlled, gold is despised, there is harmony among several religions..In a way the ideal of socialism and communism even till today.

He came out in support of very progressive ideas like divorce, euthanasia, multiplicity of religious beliefs, married and female priests in his ideal society even in that age.

Creating an alternative fictional society makes 'Utopia' a part of science fiction but Utopian literature or similar fictional societies have also been a common literary theme including Plato's Republic, Plutarch's Lycurgus, Francis Bacon's New Atlantis, Euhemerus's Panchaea among many others.

Interestingly not only Utopia is one of the pet subjects of science fiction writers but they were also enamored with the idea of its opposite dystopia or a society in which people lead wretched ,dehumanized, fearful lives particularly under totalitarian or dictatorial regimes.

It was during the period 14th to 18th century that revolutionary changes took place in development of modern science with contributions of scientists like Copernicus with his heliocentric theory, Francis Bacon coming out with the scientific method, Tycho Brahe with his astronomical findings ,Rene Descarte with his modern world view of the world as matter possessing fundamental properties and running according to certain fundamental laws, Newton with his foundations of classical mechanics, Galileo with his discovery of many celestial bodies.

These discoveries were a body blow for the myths and the ancient world view. Telescopes, steam turbines, locomotives, revolutionized the industrial landscape and enhanced the speed of human progress. This era became a fertile ground for more and more discovery and exploration and manipulation of Nature leading to thrilling inspiration for science fiction writers.

There was a change in fantasy stories about moon because moon had become the subject of scientific discovery. For example, as we got to know more about the moon with the discovery of telescope, creative writers came out with early science fiction writings with voyages to the moon where they expected to see Lunar creatures.

Astronomer Johannes Kepler's Somnium written in Latin in 1608 is considered as one of the earliest examples of science fiction where he talks about a daemon summoned by Fiolxhilde, the mother of Duracotus, a 14 year old Icelandic boy, who tells about journey to Levania or Moon.

Jonathan Swift's imaginative writings were very much part of early science fiction at a time when science itself was evolving and creating an awe in the minds of people.

His writing style was that of a person giving a first-person account by a traveller which was typical of that period and definitely provided added authenticity to whatever he wrote. He gave specific real location of the places where he found the Lilliputs and the Houyhouhms.

In Gulliver's Travels he has described bio energy experiments taking place to extract sunbeams out of cucumbers and kept those in phials to warm the air or supply sunshine at reasonable rate. Amazingly prophetic of him to think of having a price for energy consumption and that too in the early 1700s.

Mary Wollstonecraft Shelly's" Frankenstein" written way back in 1818 can be considered to be one of the first science fiction books.

The writing style was such that one felt as if it really happened. The locales are real locations in different parts of Europe.

The haunting tale of 'shame and horror' rings true as a doctor fashions a crude monster "this creature without soul" out of human body parts and it comes 'alive'.

Events don't take place on some fictional island but in real locations across Europe. The monster, far from being a fictional creature, is collection of human body parts.

The other little known work of Shelly, 'Roger Dodsworth: The Reanimated Englishman' is also science fiction and actually takes up one of the cutting edge areas of futuristic science today- cryogenics.

It narrates the story of a man who has been frozen for a long time and then comes back to life when the ice has been thawed. The story was based on a real-life newspaper report about a man involved in a "cryogenic" hoax. It is only in the late 20th century that medical science started looking at cryogenics seriously.

Frankenstein serves as a warning by depicting the result of one man's desire to play God.

This is a theme which runs recurrently in in many science fiction books and films.

Man's desire and urge to gain power over Nature must be kept under strict restraint is the moral science fiction writers have always propagated in their stories.

The power of science is such that it does need to be addressed with a sense of humility, sense of discrimination, sense of ethics and morality to be able to use it for the benefit of humankind or else it could lead to devastating results for the individual, society, and for the entire mankind.

Science and scientific inventions are a double-edged sword. They can be used for benefitting man or harming man, the choice is always with Man. One cannot blame science or scientific progress for the way Man uses the fruits of science.

Science fiction writers have dwelt a lot on these themes of how individual scientists, doctors, engineers with their highly intelligent brains have created some invention or discovered some formula for the good of mankind but then evil forces have taken it over. Sometimes the evil is not outside but just evil desire within the scientist which forces him to take the wrong path or invent machines for human destruction.

The struggle between good and evil within the protagonist or outside the protagonist, creates drama of high tension as the readers wait with bated breath for the eventual success of the good over evil.

Though there is debate whether the novel Dr Jekyll and Mr Hyde by R.L. Stevenson is a horror fantasy or science fiction. Many feel it could well be considered a science fiction because it is after drinking a potion or some kind of a medicine that there is a transformation of Dr. Jekyll into vicious criminal Mr Hyde.

The novel particularly dwells on the slow and steady deformation and disintegration of Dr. Jekyll's moral fabric and character as he is unable to control his desire and obsession to become evil and do evil deeds after changing his personality into Mr. Hyde.

Gulliver's Travels, Frankenstein, Dr Jekyll and Mr Hyde have all been adapted into movie versions for successive generations as its message of caution has a universal appeal in all ages.

French novelist and playwright Cyrano de Bergerac had established French writing in the science fiction genre in the seventeenth century itself with his "Comical History of the States and Empires of the Moon" writing about his travels to the Moon using rockets powered by firecrackers and meeting the Moon-men.

However it goes to the credit of nineteenth century French writer Jules Verne to achieve the position of being called the Father of Science Fiction along with H.G.Wells and Hugo Gernsback.

The popularity of science fiction genre can be known from the fact that Jules Verne is the second most translated writer in the world with the first position being that of Agatha Christie and third position being that of Shakespeare.

French author Jules Verne is well known throughout the world for his adventurous science fiction writings like A Voyage in a Balloon(1851),Five Weeks in a Balloon (1869), Journey to the Center of the Earth (1864), Twenty Thousand Leagues Under the Sea (1870), and Around the World in Eighty Days (1872).

The writings of Jules Verne have succeeded primarily because he has been able to convey through his stories and characters the child like excitement, curiosity and wonder felt by the early explorers/adventurers/

scientists as they went about searching for and discovering new places, creating new inventions and technologies. No wonder Verne's adventurous writings have inspired a lot of students to take up science and also become great scientists.

Again, it is the French historian, philosopher Voltaire who singularly took the creative leap in his literary work in 1752 to see earth from an alien's point of view who belonged to another Star system His short work Le Micromegas tells of a visit to Earth by Micromégas, an inhabitant of a distant planet which circles the star Sirius, and his companion hailing from Saturn . The planet from which Micromégas has come is huge, almost 22 million times greater in circumference than that of the Earth, and Micromégas is 120,000 feet tall.

Voltaire makes fun of the gargantuan ego of the humans when he points out that the aliens "nearly fell over with inextinguishable laughter" when they heard that philosophers on earth believed that the Universe was created solely for Mankind.

The juxtaposition of 120,000 feet tall Micromegas with six feet tall human itself makes the position of Man in the Universe amply clear.

This kind of science fiction satirical novel which shows the vastness of Universe (where earth may just be a blue dot) was meant hopefully to put some much required sense of humility into the mind of man.

Interestingly in 1868 world's first mechanized human or robot was conceptualized in one of America's first science fiction story written by Edward S. Ellis.

He came out with the novel 'The Steam Man of the Prairies' a fascinating idea of having a robot of a steam engine in that era.

Two men an American and an Irishman come across a huge steam powered man in the prairies. This steam-man was constructed by Johnny Brainerd, a teenaged boy, who uses the steam-man to carry him on various adventures.

The description of the steam powered man is amazing..

"Just after giving its ear-splitting screech, it turned straight towards the two men, and with the black smoke rapidly puffing from the top of its head, came tearing along at a tremendous rate."

It is further described as "It was about ten feet in height, The face was made of iron, painted a black color, with a pair of fearful eyes, and a tremendous grinning mouth. A whistle-like contrivance ......'.........door being opened in front, showed a mass of glowing coals lying in the capacious abdomen of the giant; the hissing valves in the knapsack made themselves sieve-like arrangement, such as is frequently seen on the locomotive"

It further described him as "The steam man was a frightful looking object, being painted of a glossy black, with a pair of white stripes down its legs, and with a face which was intended to be of a flesh color, but which was really a fearful red".

Another landmark work of science fiction in that era is the novel "Flatland: A Romance of Many Dimensions" written by Edwin Abbott Abbott, an English school master of London in 1884.

It describes the adventures of a two-dimensional life form, a square. To have as the protagonist a geometric figure who undertakes a journey to other dimensions makes it an extremely innovative and creative idea.

Today when 'virtual world' is becoming common place, we can understand its importance. It is only lately that scientists are talking about the possibility of other dimensions apart from our three physical dimensions. The science fiction film Interstellar does try to grapple with the issue of fifth dimension.

In Flatland the square travels through different universes which he was previously unaware of including Pointland, Lineland, and Spaceland.

The square's perspective is changed after travelling to one-dimensional, and then three-dimensional, worlds. The novel suggests that there might be higher-dimensional beings that we are unaware of which is what theoretical scientists and philosophers of today agree on.

The Time Machine written by H. G. Wells in 1895 shows time travel to a future society where people have evolved into childlike Eloi and violent Morlocks.

Again H.G. Wells called the father of science fiction was perhaps for the first time talking of time travel not as fantasy or magic but through a device or a machine trying to make it look as a real possibility.

His other work "The Invisible Man" was again a seminal work which influenced many writers and film makers. It deals with how a person is made invisible by use of optics and changing the refractive index of the body to that of air so that no light would be absorbed or reflected.

In his other novel "The War of the Worlds" which is considered a landmark he talks of alien invasion (f-Martians) which again has sparked many adaptations and many films till this day.

In 'The Island of Doctor Moreau', he takes up the whole question of Man's interference with Nature and trying to play God, when Dr Moreau creates hybrid human creatures from other animals.

Author H.G. Well's ' The World Set Free' is perhaps the best example of prophetic science fiction. Published in 1914, Wells described a new type of bomb fuelled by nuclear reactions., The bomb that he predicted would be discovered in 1933, and first detonated in 1956. Physicist Leo Szilard read the book and patented the idea. Interestingly Szilard was later directly responsible for the creation of the Manhattan Project, which led to two nuclear bombs being dropped on Japan in 1945.

In his vast portfolio of creative work science fiction writer H.G. Wells covers many areas of interest to science fiction from the extraterrestrial to time travel, to developments in human anatomy and biology, to particle physics and nuclear bomb.

Another powerful innovative and scary science fiction novel is written in 1921, "We" in Russian by Yevgeny Zamyatin where a utopia is created where the totalitarian state removes the names of people and gives them numbers and keeps strict watch on them.

In 1931, Aldous Huxley came out with Brave New World another dystopian world with bio-engineered babies, social conditioning to keep the people in a happy drugged state with 'soma', to fulfill their different roles set by the authorities.

In "1948" , George Orwell came out with his riveting masterpiece again warning of a dystopian world where a technologically advanced state wants to control freedom of its citizens by creating a nightmarish situation where your actions and movements are constantly monitored by the "Big Brother" so much so that you do not even have the freedom to think and decide on your own.

The chilling novel talks about thought crimes and Thought Police to punish individuality and independent thinking. One of the most powerful novel exposing the totalitarian regimes and mass surveillance activities, the science fiction novel is included by Time magazine on its 100 best English-language novels from 1923 to 2005. American Science fiction writer Ray Bradbury too draws a horrifying picture of a dystopian American society in his 1953 novel 'Fahrenheit 451', the temperature at which paper is supposed to burn.

The society he depicts is one where books are burnt with firemen specially tasked to do this job. Basically talking against the state censorship and attempts to control the spread of thought and ideas to the extent that scholars start memorizing books.

But it is definitely a scathing attack on any regime which attempts to throttle free speech , freedom of press, freedom to express ideas.

The subject is extremely relevant even today as most states including those which declare to liberal and democratic pay just lip service to freedom of expression and in the name of security concerns or public order or public morality, do try to control the media, have censorship on newspapers, books and films.

The novel was adapted for a film by internationally renowned director François Truffaut starring Julie Christie, Oskar Werner, and Cyril Cusack in 1966.

Sir Arthur Charles Clarke not only rode as colossus on the field of science fiction writing of the 20th century but was an inventor, undersea explorer, a futurist.

One of the landmark science fiction film of all time '2001: A Space Odyssey', in 1968, was based on ideas developed by Clarke in his short stories. Clarke wrote the screenplay of the film.

It explores the cultural and physical evolution of humans, the development of space exploration, artificial intelligence and takes up the big questions of nature of Reality, purpose of human life.

He wrote large number of successful novels including, Against the Fall of Night, Prelude to Space, Childhood's End, Earthlight, The City and the Stars, The Deep Range, A Fall of Moondust, Dolphin Island – A Story of the People of the Sea, 2001 : A Space Odyssey , The Songs of Distant Earth, The Hammer of God, from 1953 to 1997.

Isaac Asimov was one of the 'Big Three' science fiction writers of the 20th century, along with Robert A. Heinlein and Arthur C. Clarke.

He wrote and edited over 500 books both fiction and non fiction. His most famous fictional work is the "Foundation" series for which he got Hugo award where he brought in an innovative idea of psychohistory and how psychological principles could be used to assess and foresee large scale sociological and political movement combining history, sociology and statistics.

His social science fiction short novel "Nightfall" was considered to be the best short novel of all time in 1964 by the Science Fiction Writers of America.

There is an asteroid in his name besides also a crater on the planet Mars, Honda company's humanoid robot was also named after him called ASIMO. There are four literary awards named in his honor.

Asimov also may have been the first to introduce the word Robotics into the English lexicon . His short stories on robots including the famous collection , "I Robot" have really made robots and robotics familiar to most people, talking about robopsychology, interaction and relationship between robots and humans, Three Laws of Robotics, were first introduced in his short story Runaround.

Robert A. Heinlein, another of the Big Three ,science fiction writers from America is well known for his novel "Starship Troopers" which won Hugo award for best novel in the year 1960. Though considered fascistic and propagating militarism it shaped the debate about the role of military in society for many years even though depicting war between human soldiers and alien bugs.

Actually, he got his first Hugo Award in 1956 for Double Star which has the story of an actor who lands in Mars and impersonates a kidnapped politician at the brink of interplanetary war.

In 1962 his novel Stranger in a Strange Land got a Hugo award for the best novel in 1962 where a man born on Mars and raised by Martians, comes to earth. However he raised many hackles because he took up holy cows like monogamy and monotheism for attack in his book . It faced a lot of criticism and controversy.

He got another Hugo award for his 1966 novel The Moon is a Harsh Mistress about a lunar colony in the 21st century revolting against the rule of EartHe was a Naval Officer and an aeronautical engineer who emphasized the importance of scientific accuracy in his writings.

"Dune" is a 1965 science-fiction novel by American author Frank Herbert, which won Hugo award and was considered the world's best selling science fiction novel in 2003, as it showed the power struggle taking place in interstellar society to acquire a deserted area for a critical agri product "spice", mixing politics, ecology and technology. It reminds one of the competing European powers during the colonial period to get spice from India with similar mix of politics, technology and ecology.

David Lynch has made in 1984 a film adaptation of the novel. Frank Herbert's Dune and its sequel Frank Herbert's Children of Dune have also come out. A new film adaptation directed by Denis Villeneuve is scheduled to be released 2021.

Names of planets from the Dune novels have been adopted for the real-life nomenclature of plains and other features on Saturn's moon Titan.

Science fiction writer Phillip K. Dick came out in 1968 with the book " Do Androids Dream of Electric Sheep?" taking up the whole subject of artificial life, the moral implications of creation and destruction of artificial life and raising questions regarding the meaning of "life" and the meaning of "death" and whether we can justifiably apply laws made for humans to "androids".

The books served as an inspiration for the film Blade Runner.

Dick also wrote an alternative history novel where the Axis power wins instead of the Allies after World War II called 'The Man in the High Castle' which has been made into a television serial, and a science fiction novel dealing with the nature of death and reality called ' Ubik'.

British writer P.D. James's 'The Children of Men' written in 1992, is a scary dystopian novel where most of the human population has become infertile and the last man born had died.

People have comfortably resigned to their fate, with no legacy to look forward to with no children, life is losing its meaningfulness, as ablebodied population shrink migrant labour from poor countries is exploited to fulfill the need for physical labour.

The authoritarian regime (which regularly conducts mass suicides for its bored masses) takes over with no moral scruples and most of its citizens are ready to acquiesce and not raise its voice against authoritarianism as long as the regime provides "freedom from fear, freedom from want and freedom from boredom". However resistance also raises its head and baby is also born.

In 2006, a film adaptation was directed by Alfonso Cuarón, starring Julianne Moore and Clive Owen. In 2019 it was included into BBC News 100 most influential novels.

In a fairly new concept Greg Egan the author of Permutation City a 1994 science-fiction novel deals with artificial life and simulated reality.

The novel won the John W. Campbell Award for the best science-fiction novel of the year in 1995 and was nominated for the Philip K. Dick Award the same year.

It explores the fundamental question in today's world where virtual reality is slowly gaining ground as to how can we know that a human is not computer simulation but a "real" person. The question which is also dealt with in Matrix. Can consciousness be produced by computer program ?

Egan's story deals with the life of the copies or digital renderings of human brains with complete subjective consciousness, living within virtual reality environment. Their life and assets depend on how much computing power they have or can garner.

Knowing one is a Copy, a computer simulation previously recorded from his physical self, can lead to innumerable existential questions regarding self-consciousness, identity, reality.

In 2011, Andy Weir wrote, "The Martian"', a science fiction novel about an American astronaut, Mark Watney, who is stranded on Mars and has to survive till he is rescued after 18 months. Son of a particle physicist and electrical engineer, Weir who himself also has a background in computer science did a lot of research in areas of orbital mechanics, astronomy, spaceflight for writing a very realistic book.

The paper back edition of the book remained continuously at the number one position for 12 weeks on the New York Times Best Seller List. The Wall Street Journal called the book "the best pure sci-fi novel in years"

This book is a rare study of Man's determination, powerful survival instinct , skill and intelligence with tremendous amount of sense of humour and optimism while alone in a remote planet in space for nearly 18 months. The book was adapted into the film 'The Martian', in 2015, directed by Ridley Scott and starring Matt Damon.

CHAPTER 2
DIFFERENCE BETWEEN SCIENCE FICTION AND FANTASY

There is a fine distinction between fantasy and science fiction and that is science and logic.

In fantasy an author's imagination is the only constraint on the story line, the characters, and the happenings within a novel.

In science fiction the boundless flights of imagination have one powerful check in the shape of the scientific laws and principles, scientific developments,and discoveries.

One of the fundamental difference between science fiction and fantasy is science fiction narratives describe what is possible whereas fantasy narratives describe what is impossible.

The world of science fiction has to be firmly based on reality even though it may relate to a civilization in the other part of the universe or a time period millions of years ahead.

In contrast, fantasy elements are founded purely on facets of the imagination.

The story in a science fiction novel can speculate from the present status of scientific development in a particular field. It can just takeoff and play around with all the possibilities emerging out of that scientific development or discovery, in the coming years ahead. The connection may seem tenuous at times, but it must still be firmly grounded on today's reality regarding the scientific invention, discovery, or principle.

Secondly and more importantly the author cannot throw away basic logic and rationality when dealing with the characters and events in the novel.

Renowned science fiction writer Isaac Asimov has said science is based on facts and it exists in reality. It can be about robots, aliens, space, etc.

However, fantasy is just imaginary. It has no connection with facts. In science fiction, we may discuss the things that can happen in real time, but in fantasy we can and do talk about supernatural things which do not exist in the world. Events can happen at spur of the moment without any logical reason which may not be possible in science fiction.

Fantasy has been a part of storytelling even before the advent of written literature, it has no connection with science, whatever man can imagine or think of can be part of his story.

Technology plays no role in fantasy though it has an important role to play in science fiction.

Both science fiction and fantasy however answer the very important question "what if ?" They both build hypothetical worlds in answer to this question.

What if aliens land on earth? What if androids become conscious? What if teleportation would be possible? What if trees could talk? What if one could get one's wish by wave of the wand?

It is to answer such questions that both the science fiction writer and the fantasy writer puts his pen to paper.

They both write speculative novels but while science fiction has to be more realistic and reasonable while extrapolating from scientific developments, the fantasy novel can be purely imaginative.

Science fiction because it is tied to science and technology its speculations are limited to the possible and probable, however fantasy novel can transcend any such limits it can visualize things which are impossible or "fantastic".

Magic is a key element in fantasy which helps the characters achieve all their goals which is almost taboo in science fiction. Magic serves as the single point answer for all difficult questions in a fantasy.

In fantasy you will find prophecies coming out true, you will find enchantments and charms having a powerful role to play, all without any rational basis but pure magic. However, fantasy also does not mean complete lassie fair. The fantasy author's world too has to follow certain rules though they may all be created by the author himself.

Magic may play an important role but author cannot use it as a convenient plot device every time he needs to look for a cause or motivation or wants to get things done.

The magic system must have its own established rules and powers. When, where, how and who can use magic is very important, so even if it is the world of fantasy the credulity of the reader is not tested. The basic normalcy and reasonableness have to be followed even if the world is completely imaginary, otherwise even the readers will not be able to understand what is going on.

While sci-fi can take place in a world resembling ours, the fantasy world where magic rules is completely new and a different world built by the author.

World building is an important aspect of the fantasy writer. Like C.S. Lewis has built up an entire fantasy world of Narnia or J.R.R. Tolkein has built in Lord of the Rings.

Once we as readers accept the notion of the entire fantasy world being a world full of magic, we need not be surprised or be amazed at the fantastic happenings within the world.

The dictionary meaning of fantasy is "a book, movie, etc., that tells a story about things that happen in an imaginary world."

The more the depth and finer details (the people, the language, the society, the traditions, and history) with which the writer is able to build his fantasy world the more comfortable and at ease is the reader with the display of magic which comes naturally as part of it.

The science fiction writers too actually build worlds which maybe utopian or dystopian world, a world in a different part of the Universe, a world in a completely different time period , a world in a different dimension. However, they must have to adhere to basic scientific laws or logically link the story to the science from which they are extrapolating. The reader must get the sense that whatever is happening is within the realm of possibility.

Fantasy always deals with supernatural or non-existing characters which are not seen in their real world. It is the main reason Star wars is recognized as a fantasy rather than a science fiction.

Actually Lord of the Rings is a fantasy story and Star Trek is a science fiction story . In Lord of the Rings its author J.R.R. Tolkien builds a detailed world full of elves and trolls and hobbits. It' s a magical world where once we suspend our disbelief we enjoy the happenings as if it is happening in the real world. He includes superior races but also brings in humansso that the readers have a connection. But as a genre it is a complete fantasy.

Science Fiction deals with a wide array of alien races, and creatures that maybe unlike anything seen in our world, however we accept it because they are all not of this world and science has not yet ruled out that aliens do not or cannot exist. Its up to the imagination of the science fiction writer as to the type of alien he wants to create as long as the being is even remotely possible according to our present scientific understanding.

Science fiction can therefore show aliens and their world. They can be composed of substances completely different than found in our own world like silicon or methane. However, one thing is sacrosanct even when creating those worlds and alien species basic principles of science, physics, chemistry, biology must be kept in mind. Events and actions cannot take place just magically without a proper reasonable and logically valid explanation.

Science fiction deals with scenarios and technology that are possible or maybe possible based on science. Some science fiction such as time travel stories may seem implausible, but they are still not beyond the realm of scientific theory.

On the other hand, fantasy generally deals with supernatural and magical occurrences that have no basis in science at all.

Fantasy is the oldest genre of storytelling as man must have at that point of time completely depended on his imagination to make sense of the world. So, we have imaginary stories regarding almost all the natural objects and events we see around us including animals and plants, light, darkness, sun and moon, mountain and sea.

In Indian folk tales we have the well known tales of "Singhasan Battisi" or "Thirty two tales of the Throne" . The throne of ancient king Vikramaditya is found by King Bhoja which has 32 statues of apsaras that had been turned into stone due to a curse. Each of the apsara tells Bhoja a story about the life and adventures of Vikramaditya, in order to convince him that he is not deserving of Vikramaditya's throne.

This is classic fantasy story with elements of magic. Well known collection of stories like Panchatantra or Arabian Night and Aesops or Singhasan Battisi fables are all fantasy which mix personified animals,supernatural events with real life and usually impart moral lessons.. Homer's "Odyssey," the Arthurian Legend (the stories of King Arthur) are all fantasy.

Elements of fantasy like supernatural beings, monstrous beasts, magic abound in ancient mythologies, folklore, and religious texts around the globe But today fantasy is slowly coming into mainstream as a Literary genre.

One of the early authors to write fantasy for the adults was Scottish author George MacDonald, who wrote Phantastes in 1858 about adventurous of a young man in the dream. . William Morris, an English author created a whole new world for his fantasy novel "The Well at the World's End " in 1896. The Lord of Rings by J.R.R.Tolkien was written around 1954 again depicting an elaborate fantastic world and making fantasy a popular genre.

Around 1956 came C.S.Lewis's T he Chronicles of Narnia series, Terry Brooks' "The Sword of Shannara"(1977) that went ahead to become bestseller.

It was J. K.Rowling's Harry Potter novels which made fantasy and magic almost household words world over with movie adaptations and television shows.

Superman who can move faster than a speeding bullet and can fly like a jet ,is just a fantasy character and not part of science fiction.

This is because his superpowers are not based on science atleast the science as we know it at present. Superman without wearing any bullet proof vest is bullet proof, and what is the mechanism which gives him the ability to fly we do not know.

Actually Superman has been shown to be most powerful as well as invincible person except perhaps the radiation of kryptonite the green crystalline material which takes away all his powers.

Superman can move entire planets, can fly, can travel faster than light and reach other planets and galaxies within minutes, can with stand atomic explosion and fly through the core of a star. He has X ray vision and heat vision and can see across interstellar distances and observe events that occur on a microscopic and even atomic level. He can also see across the full electromagnetic spectrum, including infra-red and ultraviolet light.

He is able to do all this by breaking all possible scientific laws of physics. This is precisely why he cannot be a character of science fiction. He is and will always remain part of the fantasy world which all children and many adults will continue to love.

It is because Superman belongs to the world of fantasy we do not question as to how and where he got all his powers from. The answer that he has his powers only being an alien from Krypton is clearly not satisfactory to our rational mind but as a fantasy story character we unquestionable and thoroughly admire his super powers .

Another extremely popular series "Star Wars" " too cannot stand the rigid scrutiny of being a science fiction novel though many of its fans will like it to be one. Star Wars seems to many to be science fiction because of space battles and robots but then George Lucas has himself said that "Star wars isn't a science fiction film its a fantasy film and space opera."

In Star Wars there is no explanation given for the Force which is the crucial element playing a pivotal role in granting power to different characters at different times like psychic abilities such as telekinesis, mind control, and extrasensory perception. The Force is sometimes referred to in terms of "dark" and "light" sides, with villains like the Sith drawing on the dark side to act aggressively while the Jedi use the light side for defense and peace.

In science fiction, your creations need to make sense within the natural laws of the universe (our universe.)

To understand the difference between science fiction and fantasy we must see the reasons why despite all its extraordinary characters and events The Matrix is considered to be a science fiction.

The reason clearly is that the hero Neo's actions do not happen because of some magic neither the author gives any kind of paranormal solution.

The author builds a full virtual world (the possibility of which cannot be denied even scientifically today) for Neo. The entire population of earth is said to be living their whole life in this neurally based interactive simulation like a virtual reality video game.

Once the simulated world is established and if Neo is in control of that then Neo can fly, stop bullets in mid air and move at superhuman speed because it is all simulation and thus Neo is not breaking any scientific law.

It is science fiction but it is not fantasy because the author has fulfilled the fundamental criteria of basing the story on latest scientific developments in the field of artificial intelligence, virtual reality.

Science fiction thus needs scientific explanations whereas fantasy does not need scientific explanations for why one character can fly, while another breathes fire they are all accepted as we are in the world of fantasy.

From the world of Super Heroes you have Batman who is part of science fiction as his super powers are based logically on high-tech gadgets and his physical powers are due to his own training in martial arts and his powers of detection developed by his own technology. He does not flout natural laws and therefore does not demonstrate the kind of powers several others do. He is believable. But Spider Man getting his powers being bitten by a radioactive spider does pass muster as hard core science fiction and would primarily remain as a fantasy.

While science fiction stories often take place in a dystopian, hyper-technological future. Fantasy stories are traditionally set in worlds populated by mythical creatures and supernatural events.

Stories of Superheroes are not considered science fiction because even though they live in a world that looks very similar to our own, but their powers are sometimes supernatural with no logical, scientific explanation.

Iron Man is very much part of science fiction even though he performs like a superhero. This is again because all his powers are drawn from highly sophisticated gadgets.

Stark besides being a top industrialist is a scientist and technologists to be able to build himself a suit of armour that gives him superhuman powers.

No such suit may exist but work on such wearable exoskeleton is also being undertaken by scientific establishments in the world today.

He has the ability to fly but that is not because he just wills to fly but due to a rocket belt pack.

Technologies used by Iron Man are also taken from the latest cutting edge technologies which are slowly being made available like three dimensional tactile interface which he uses to design his armour, is also based on new technology which is slowly becoming available.

Science fiction must see that despite all the superior advances in science and technology, the world still feels believable. There are no supernatural elements that defy the laws of the universe. All of the technological aspects have to follow a convincing line of reasoning.

j Proper compliance with the laws of physics can be a useful tool for classifying science fiction versus fantasy.

CHAPTER 3
SCIENCE FICTION AND PHILOSOPHY

Science fiction films delve into the philosophical questions regarding the nature of existence, the nature of Reality, the destiny of Man, the power of Nature, under the veneer of science and imagination.

Science fiction literature helps us bring into focus the abstract ideas with which we deal to explore some of the most fundamental philosophical problems facing humankind, particularly the nature of the world in which we live.

It is primarily because of its underlying search for answers to some of the universal questions that science fiction films have had a remarkable hold on the imagination of filmgoers all over the world.

The developments of modern science and technology brought about a schism between Man and Nature, Man and Spirit almost permanently. He was divorced and cut off from Nature and his own Higher Self.

Instruments like the microscope and the telescope ripped open the world of Nature which had fascinated Man and his imagination since the advent of Man on earth.

He was able to see the inner workings of all things around him, the animate and the inanimate including plants, animals, and humans. The telescope showed him that his world is no longer the centre of the universe as he believed but a speck of dust in the vastness of space with millions and trillions of stars and planets.

The fruits of knowledge had its downside, Man lost his innocence, his element of surprise, his awe and amazement at the wonders of Nature.

The gods, the fairies, the devil and all the other fanciful ideas and myths of Man which he had developed over centuries upon centuries to explain the natural phenomena and Man's role in the world was gone.

Scientists probing into the depths of Nature sought and came out with scientific laws which laid bare the principles on which the natural world functioned.

These could be universally replicated once one knew the scientific law or principle behind it.

Science and scientific laws revolutionized human progress and civilization bringing about a quantum jump in the speed of development. With rapid growth in number of inventions, changes which took 100 years earlier started happening in just 10 years telescoping the entire process and making it virtually impossible for humans to cope up with the changes.

Man, himself had changed from a species interdependent on other species in a wide network, under benevolent Nature, to an exploiter and manipulator of Nature and abuser of other species.

Scientific advances helped make Man's position much stronger in the hierarchy of animals and other life on the earth. The large size and complex brain led this highly intelligent primate to invent all possible gadgets to conquer earth and enslave all other species in the relentless march of human progress, development, and prosperity.

The other close relations chimpanzees, gorillas, Orangutans were left far far behind.

It is only in the late 20th century that there was a horrible realization by Man that merciless wiping out of other life forms in the march of human Progress will result in environmental degradation and ecological imbalance and eventually lead to the extinction of Man himself. Efforts though belatedly and in extremely small measure then started towards protection of Nature and safeguarding of animal and plant life.

Developments in astronomy had already dislodged Man and his world from being the Centre of the Universe to being just an insignificant part of it, theory of Evolution further eroded Man's ego when it claimed that Man instead of being made in the image of God had evolved from apes.

As science progressed Harvey laid out the details of blood circulation within the body and John Dalton came out with the atomic theory of matter. Electricity and magnetism were discovered and found to be forces of nature which man soon could control and utilize for his development and progress.

Man was no more in the fear of the gods who were supposed to control the forces of Nature rather becoming his own Master and of all that he surveyed.

These were developments which fundamentally changed Man's view of himself. The march of science continued and challenged even the most holy cow of human society –Religion. Man, soon started questioning God and playing God.

It was his belief in God and religion which had always given solace when Man sought answers to fundamental questions like Does Man have a soul? Where does Man go after he dies? Is there God? Who created the Universe? Does Man have free will?

The faithful placed all questions to which they had no answers at the feet of the Almighty God, the eternal being, the Supreme power, the great Unknown since time immemorial.

The impact of science and technology was that while unsettling the faith of the common man on Supreme Power it raised more questions than it had answers to.

Science which was supposed to give answers to the critical fundamental questions of humankind was not equipped to answer those and they continued to trouble the thinking Man.

In this scenario the sensitive creative mind naturally give rise to speculative science fiction as one of the avenues to raise issues for which science did not have a ready answer?

Science fiction is that branch of literature which deals with – What has been and what can be the impact of science and technology, the inventions, and discoveries of today and possibly of the years ahead at the individual and societal level. Speculation on

What If ?has been the main inspirational drive for many a science fiction book and film.

With the first rockets being built and space flights becoming a reality and Man reaching the moon? Man's imagination could no longer be confined to the earth and science fiction too took flight and went all over the galaxy and beyond.

No wonder encounters with aliens, alien invasion and alien civilizations in all corners of the Universe of all varieties both friendly and hostile became the staple fare of Science fiction.

Despite it being science fiction fantasy, today the possibility cannot be denied that someday we may contact aliens, it touches deep chords in the human mind because once we have gained the ability to go beyond the earth and reach for the stars, the alien encounter whether on earth or some other planet becomes an inevitability with only a question of time.

Naturally no science fiction writer would believe that we are alone in the universe, as it goes completely against the grain of human imagination and logic.

Space flight, voyages to different planets and universes, finding new civilizations and new kinds of aliens, new sources of energy, aliens arriving on earth and being countered by earthlings if their intentions are malicious have routinely become part of science fiction.

To the imaginative and speculative science fiction writer the field is wide open to come out with all kinds of exciting and adventurous, thoughtful, inventive, innovative stories and novels as long as he bases his story on scientific developments and does not go against basic scientific laws in his flight of fancy.

Apart from space flight the next big technological revolution was the invention of micro chip and computers which led to the advent of digital world which completely transformed the entire human scenario.

Perhaps for the first time in human history there was even a remote possibility of humans being completely replaced by machines.

When brain started being looked upon as a computer by scientists came hosts of fundamental questions as to the nature of Man itself. Are we just machines? Do we have a soul? Can computers act as humans? Can they have emotions?

These scientific developments seem to have just brought into question the whole issue of_ What does it mean to be human?

A number of sci-fi films seem to be dealing with just this question. Two of the major films being_ I Robot and _Bicentennial Man'

With supercomputing, the power of computers was enhanced manifold. The growth in the field of cybernetics, artificial intelligence and robotics with sensors mimicking eyes,ears and hands, the gap between man and machine was also slowly and surely being reduced.

The human-machine interface which has become the key form of engagement in the 21stcentury world has spawned innumerable sci-fi movies. Science fiction books and movies became the platform on which the philosophical questions about personal identlty, moral agency, artificial consciousness and what exactly it means to be human was being raised again and again.

With science fiction genre being open to flights of fancy science fiction writers could come out with alternative scenarios as to what could be the consequences of the development of certain technologies like informatiln technology on human society and environment.

Suppose intelligent robots become a substantial portion of our workforce (a very likely scenario in future) what will be our reaction? How will we treat a robot? What will be the difference between an intelligent robot and an intelligent Man - emotions? If a robot shows emotions. Then will it become human? Will a robot be able to murder intentionally?

These are the kind of fundamental questions science fiction can and does raise. The story of ' "I Robot " the film deals with these kind of questions.

In a particularly poignant moment in the film— I'I Robot' the robot Sonny shows human emotion in flashes of anger and also seems to demonstrate human understanding when he displays a typically human gesture of_ ' winking' which humans do when they share a secret or joke with another.

Sonny clearly displays the inner spark of a human and takes actions as a self- motivated agent which goes against the mould of his character as a machine.

As we get to know Sonny the robot better we find that it is self aware, afraid of death,has dreams and shows signs of morality. He acts on his own disregarding the laws meant for Robots at his own discretion. Much to our shock even though a machine he seems to follow his own inner logic.

On almost the same theme Isaac Asimov had made the powerful film_"The Bicentennial Man 'where the robot Andrew almost forces humankind to declare him as a human for which he is even ready to die. At the end he actually dies if one would accept the concept of _'death' of a machine.

The film narrates the story of Andrew, who from being the Martin family's helper in home maintenance as a robot slowly becomes a very close member of their family demonstrating many human qualities like sentimentality, creativity, humour.

His _ "personality" is not part of genetics

neither part of his hardware (he is a machine) .

Philosophically it is one of the most enduring films where the whole question of_ ""What makes us human?" is dealt with and can be seen in automaton Andrew's attempts to become human.

In his desperation to become fully human he upgrades and provides himself with enhanced facial expression and human-like skin and also a central nervous system. He gets himself operated to add prosthetic organs in his body like an artificial liver, kidney, and even stomach.

Its almost painful and touching to see the efforts the inanimate object Andrew takes to become human. He also loves and lives with Portia and has sex.

To be recognized as fully human he is even willing to pay the ultimate price—death. He gets a neural system and circulating blood to make him age over time.

It is almost pathetic to see as he tries his best to become human. Andrew lives on, as other members of the family he knows die one after another, as time passes by in the natural course. Being a robot he cannot die.

Death is an important and crucial element that makes us human. All our lives are temporary and does not last forever. But that exactly is what gives the zest, the excitement to our lives and to us, it is what the sensation of living means, it is what is to be human to be— alive.. The— life in us makes us human. Once that pulsation stops and life is gone, we are dead like any other inanimate object or machine.

Machines do not have life…they will not die but at the same time they do not have the experience of life or being ve.t is a strange dichotomy. Andrew is ready to die to sacrifice his eternal existence to experience truly what it is to be a human.

Author Hannah Kaufman talks about the central character, a robot, in the film Ex Machina by Alex Gardner "You have the female robot Ava, perfectly sculpted face with the round brown eyes and light skin tone of a Caucasian woman, and the outline of her body is the epitome of feminine beauty. Yet bits and pieces of this body are wrapped in high-tech mesh or are transparent enough to reveal the galaxy of electrical wires and gears churning below the surface, serving as a constant reminder of her mechanical nature.'

She is the object of love or hate? Can we love a machine more than a human? Can machine know what is love? Can machine show emotions of love, hate, jealousy? Do questions of love, guilt, conscience trouble machines or robots? Are they like humans? If not, what is the difference? What should be our relationship with robots? These are philosophical questions and again go to the root of what makes us human.

Another extremely important basic epistemological question, — how do we know that the world around us is real? How do we know what we know and that it is true?

Philosophers have been always grappling with this problem because our senses are limited and can only tell us about the sensory phenomenon which we see all around us, but not about the noumenon or the Reality which lies behind it. For example our human hearing may be limited to certain fixed range of frequencies that does not mean that audio frequencies in reality consist of only those many frequencies. We are limited to a three-dimensional world but that does not mean more dimensions are not there.

Similarly, both our vision and touch are also limited and there may be many things existing in the world which we are neither able to see or touchWell known philosopher Descartes makes the point very clear when he puts into doubt every knowledge which he gets through his senses.

Descartes concludes that he cannot rely on his senses, and that for all he knows, he and the rest of the world might all be under the control of an evil demon.

He wonders if the God he has always believed in might actually be a malevolent Demon capable of using his omnipotence to deceive us even about our own thoughts or our own existence.

Thus, there is nothing in all his experience and knowledge that Descartes cannot call into doubt . He says after examining all the things he thought he knew about himself and the world, he concluded that the only thing he knew with absolute certainty is that ''_I am, I exist''. Because if he denied his own existence and put it also into doubt, then that idea or thought itself will also confirm his existence. Therefore, it is the only certainty.

Like Descarte's evil demon in the science fiction film Matrix the artificial intelligence forces a virtual reality on humans. Just as Descartes realized that the sensations in his dreams were vivid enough to convince him the dreams were real, the humans who are plugged into the Matrix have no idea that their sensations are false, created artificially instead of arising from actual experiences.

Until Neo is pulled out from the Matrix, he, too, has no idea that his life is a virtual reality. Like Descartes, Neo eventually knows to take nothing at face value, and to question the existence of even things such as chairs and tables that seem most real.

In Matrix the computer-generated dream world was designed by an artificial intelligence which utilizes human bodies for extracting energy while distracting them via a relatively pleasant parallel reality called the— M= Matrix.

How do we know that what we think is real is real? Morpheus asks Neo—Have you ever had a dream that you were sure was real? How would you know the difference between the dream world and the real world?''

Ancient Chinese Toaist philosopher called Chuang Tzu one night went to sleep and dreamt that he was a butterfly. He dreamt that he was flying around from flower to flower quite sure that he was a butterfly. But when he awoke, he realised that he had just been dreaming, and that he was really Chuang Tzu dreaming he was a butterfly. But then Chuang Tzu asked himself the following question: "Was I Chuang Tzu dreaming, I was a butterfly or am I now really a butterfly dreaming that I am Chuang Tzu?"It is also very similar to the Indian Hindu concept of Maya in Advaita school of Vedanta which states that this world is like a dream world created for the humans which is not the actual Reality.

In Hindu philosophy Maya is the illusory power of the Supreme Being which creates the whole world in which we are all entrapped. We see ourselves as individual beings in the three dimensional world and under Maya's delusion we undergo the cycle of birth, death, rebirth. We fail to discriminate between truth and false hood till we recognize our true divine self. It is only eons of struggle and with discrimination, dispassion, longing for liberation from the world of Maya, by controlling the mind and senses, and as devotees point out with the grace of the Supreme Being that finally one can cut asunder the veil of Maya and realize his own real Self.

Enlightenment is achieved when one awakens from the illusory world (maya) and becomes one with the Bramhan, the absolute, which is the truth.

While Neo wants to extricate himself from the simulated world ,Vedanta urges men to realize that the world is Maya or like a dream and make all efforts during this human life to wake up to Reality or achieve realisation.

In similar lines Greek philosopher Plato too talks about shadows cast on the cave's wall. Plato explores the idea that the real world is an illusion in the allegory of the cave in The Republic.

Plato imagines a cave in which people have been kept prisoner since birth. These people are bound in such a way that they can look only straight ahead, not behind them or to the side.

On the wall in front of them, they can see flickering shadows in the shape of people, trees, and animals. Because these images are all they've ever seen, they believe these images constitute the real world. One day, a prisoner escapes his bonds. He looks behind him and sees that what he thought was the real world is an elaborate set of shadows, which free people create with statues and the light from a fire.

Renowned philosopher David Chalmers wrote in a paper for the philosophy section of the official The Matrix website ((2003) that the Matrix presents a new version of an old philosophical fable: the brain in a vat.

According this philosophical argument a disembodied brain is floating in a vat, inside a scientist's laboratory. The scientist has arranged that the brain will be stimulated with the same sort of inputs that a normal embodied brain receives.

To do this, the brain is connected to a giant computer Simulation of a world. The simulation determines which inputs the brain receives. When the brain produces outputs, these are fed back into the simulation. The internal state of the brain is just like that of a normal brain, despite the fact that it lacks a body. From the brain's point of view, things seem very much as they seem to you and me but it is all false.

It has all sorts of false beliefs about the world. It believes that it has a body, but it has no body. It believes that it is walking outside in the sunlight, but in fact it is inside a dark lab. It believes it is one place, when in fact it maybe somewhere quite different. Perhaps it thinks it is in Tucson, when it is actually in Australia, or even in outer space. Neo's situation at the beginning of The Matrix is something like this , says Chalmers.

He thinks that he lives in a city, he thinks that he has hair, he thinks it is 1999, and he thinks that it is sunny outside. In reality, he is floating in space, he has no hair, the year is around 2199, and the world has been darkened by war.

There are a few small differences from the vat scenario above: Neo's brain is located in a body, and the computer simulation is controlled by machines rather than by a scientist. But the essential details are much the same. In effect, Neo is a brain in a vat.

A film like Matrix raises philosophical questions says Chalmers like— Hoii w do I know that I am not in a matrix? After all, there could be a brain in a vat structured exactly like my brain, hooked up to a matrix, with experiences indistinguishable from those I am having now. From the inside, there is no way to tell for sure that I am not in the situation of the brain in a vat. So it seems that there is no way to know for sure that I am not in a Matrix.

Oxford philosopher Nick Bostrom too has come out with the theory that the world we are all part of may be a simulation created by humans many centuries ahead in the future when it would be possible to create such a large scale simulation.

Man machine relationships is one issue but another pertinent issue with philosophical ramifications underlining what makes us human is man-animal relationship.

One of the best and most disturbing science fiction film dealing with this issue is Planet of the Apesa film based on the novel by Pierre Boulle.

The film released in 1968 completely stunned the audiences. The scenes were awesome and scary, human beings cowering in fear and running for their lives amid fields of crop like rats while apes rode on horses hunting them down or capturing them.

If humans are ever treated like the way we treat many animals even today, just the thought is likely to be a nightmare and will leave us with goose bumps?

Think for a minute we eat goat meat and there are mutton shops? They haveat carcasses hanging openly for everyone to view and the shop keeper is ready to cut pieces of your choice. If the scene is reversed it can be a horrible experience to watch human carcasses hanging in a meat shop i In a goat world.

In a parallel universe where animals are the dominant species and treat us like the human race treats animals would really be scary as some satirical caricatures have shown. There is one which shows a tiger sitting on the sofa reading a newspaper with number of busts of hunters hanging on the wall as prized collection, the rug on the floor is actually a full skinned hunter his with head, like a rug of tiger's pelt and head.

It is science fiction which gives us glimpse of this kind of a reversal of situation in Planet of the Apes, where human kind has lost its position of preeminence.

In the prequels to the first movie came others which just show how the Apes slowly take over the world.

Human beings being chained and caged, humans being looked down upon as inferior or despised as a species by Apes is something which goes against our basic instincts. Even apes showing us pity and kindness is something which is difficult for us to swallow.

Should animals have same rights? Are they as intelligent as humans ?Are they self conscious like humans? Do they have a conscience? What is ethically the correct relationship of man and animal? Can Apes enslave men ?

These are issue which really trouble us as we see the rise of the Apes in these series of movies.

Of course, there are innumerable other science fiction movies which show Man confronting both dangerous and large sized animals, birds, fishes and insects. Interestingly other movies showing animals or birds or insects taking over the planet does not create the kind of fear which planet of the Apes does because we do not take those possibilities seriously.

The lack of scientific understanding of genetics in the 19th century did not prevent science fiction works such as Mary Shelley's 1818 novel_ Frankenstein' and H. G.Wells's 1896 novel _""The Island of Dr Moreau""' from exploring themes of biological experiment, mutation, man-animal hybridisation, with their disastrous consequences, asking serious questions about the nature of humanity and responsibility of science.

The knowledge about the Human Genome and advances in genetic engineering has again brought forth a plethora of questions in front of the humankind of ethical and philosophical significance with which we are still struggling.

These are areas which till today were left primarily in the hands of God or fate.

With genetic engineering we are making fundamental changes in Nature of the individual or the species. We are playing God.

Should man act as God? We could try to make the —perfecthuman being but would that be the most advantageous for the individual ? Do we control all the other circumstances of the individual in his life? Maybe little natural imperfections will make him or her handle his/her circumstances better. How do we know? We cannot foresee the future. Do we control destiny?

These philosophical questions are again raised by science fiction films from Mary Shelly's Frankenstein to Gattaca. Human tinkering in genetics can have widespread and far- reaching consequences of which we do not have any idea or any kind of control.

The morality of creating genetically engineered humans, and the ability to act contrary to our biological predispositions are issues raised in this extremely relevant movie.

An advertising line for the movie: —"" There's no gene for the human spirit""correctly points out that human spirit is free and cannot be under any kind of a deterministic gene or fate.

Issues like acquiring enhanced skills incorporated in the child by paying higher price brings in whole new area of concern. How will genetic profiling be used by insurance companies?

Gattaca raises the question of whether the most important factor in determining a person's fate is their genetic predisposition or the force of human will and determination which science cannot account for.

One of the best of science fiction films of all times and pioneering in its philosophical sweep is Stanley Kubrick's 2001: A Space Odyssey.

The space flight of the pair of astronauts on a voyage to Jupiter with a sentient AI named HAL 9000, very soon becomes a symbol of Man's spiritual journey towards enlightenment.

The 1968 film when man had not even reached the Moon, beautifully depicted the vastness of space and Man's desire and struggle to reach the boundaries of the Universe.

The 2001: A Space Odyssey really takes on the big question of Man and his destiny. Is there a meaning to life? As the film seems to take on the entire gamut of evolution of humans from Ape age to becoming a— star child. It is the ultimate philosophical film with innumerable interpretations. It is still an enigma for the viewers.

Time travel also raises various ethical and philosophical Issues of determinism and freewill. How human Intervention in the past can change what is going to happen in the future? Is it right? Would it lead to the desired result or something else? How would knowledge of the future affect your present? Can you really change your future or your destiny?

Science fiction film Minority Report deals with Time travel and issues of predestination and free will with its Plot about three people capable of precognition that allows them to know when someone is about to commit murder. But can we know for sure that someone will commit murder? Do we believe they have free will or not? If we have no freewill, are we then even morally responsible for our actions?

Rarely do science fiction films take up dry serious topics like language. One such movie is "Arrival" where the linguist Louise Banks (played by actress Amy Adams) is the main protagonist and is tasked with deciphering alien script and language.

The entire philosophy of language deals with human communication and how it takes place , what kind of impact does language have upon human understanding and development.

In this context one must understand the crucial importance interpretation of "language" whether audio or written symbols may have if ever humans were to be in real contact with aliens.

How exactly will we be able to exchange information about ourselves and acquire information about the aliens. What we call our innumerable languages and scripts will be understandable to aliens or they will look upon it as symbolic gibberish. Our grand speeches may not convey anything apart from incoherent sound to the aliens. A truly despairing and frustrating thought.

In the movie Arrival we see Louise Banks while she is busy understanding the language of aliens because the aliens perceive time in a circular fashion and not in a linear fashion, losing her linear sense of time and have visions of past and future more frequently.

Basically Arrival gives credence to the controversial Sapir-Whorf hypothesis which states that language is not just a tool to express thoughts but does influence the way we think and see the world.

The main argument was that in different cultures the meaning changed of the same word. Like in some cultures the word red would mean just red but another culture may have names for hundred different shades of red. Therefore witnessing a red object depending on the shade they would be naming it and perceiving it differently like cardinal or madder, while most other culture would just see and say red.

Basically Sapir-Whorf hypothesis calls for " linguistic relativity" or that our perception of the world is shaped by our native language.

The movie Arrival to some extent breaks new ground by focussing on the complex and fascinating area of language and highlights to what extent language influences thought and thought influences language debated for centuries by philosophy of language and philosophy of mind.

Another important issue which many science fiction films deal with is Cloning which raises a number of religio-ethical and philosophical questions.

There are

issues which come up like when does the embryo becomes a "person" ? Is destroying an embryo murder? Many countries have prohibited human cloning and call it crime against human species. One of the worries has been that early implantation of these cloned human embryos always results in their death and often causes maternal death or morbidity. There is also a view that it leads to serious abnormalities and birth defects.

The whole concept of accepting children with humility as gifts rather than possessions or trying to create with help of technologies like cloning "designer children" of your own choice and liking for your own happiness is abhorrent and immoral and against basic human rights of the child.

Child is not a commodity or a "made to order" product and you as parents cannot decide their traits according to your whims and fancies.

CHAPTER 4
SCIENCE FICTION FILMS

Science Fiction films are a genre which cuts across the age barrier and is loved by almost all lovers of adventure, fantasy and science from nine year olds to the ninety year olds.

The Sci-fi films have a history almost dating back to the beginnings of cinema itself. Even in the silent film and black and white era we have had some remarkable science fiction films.

Though a lot of total fantasy films are called science fiction they do not fulfill the very basic premise of science fiction films that they must be based on or must draw their inspiration from some basic scientific principle or invention.

The bedrock on which a sci - fi film stands is that it has to be scientific and logical. The story must be in the realm of possibility and probability rather than being a purely imaginative story (fairy tales, magical stories, mythologies) which is logically impossible in real life.

As far as science fiction films are concerned if we want to take a film based on scientific development then perhaps 'The Mechanical Butcher' released in 1895 'La Charcuterie Mécanique' made by Lumière Brothers could perhaps be one of the first science fiction film .

In the early years of cinema this short film just showed a live pig being turned into pork products with help of a machine.

Using simply a jump cut (A jump cut removes the footage between two points to delete a period of time) the film depicted a live pig being placed inside the machine at one end of the stand and finished products being taken out of the other end of the stand, much to the surprise and delight of the viewers.

Renowned French director Georges Méliès came out with the silent film 'Gugusse and the Automaton' in 1897 where he showed the amazement of a clown by the mechanical movement of an automaton. The film marked the first known cinematic appearance of a robot.

In Britain a short silent science fiction film 'The X-Rays' came out in 1897. It showed how the new scientific gadget X ray could reveal the interior of a person and an umbrella.

In one shot you have a fully dressed gentleman and lady with umbrella. They then get exposed to X-Ray. The following shot has a skeleton with the actor wearing black dress and bones marked out in white paint and a lady with an umbrella with only the spokes on it. Then you had them come out of the machine and they were again back to normal being fully dressed with proper umbrella.

The film was directed by George Albert Smith. This film again shows the very early use of jump cuts to show the transformation of the couple after being exposed to X Rays.

Georges Méliès made 'Under the Seas' in 1907 a parody of the 1870 novel 'Twenty Thousand Leagues Under the Sea' by Jules Verne, depicting a fisherman's dream of taking a submarine to bottom of the ocean, where he sees shipwrecks ,sea nymphs, sea monsters, starfish, mermaids.

In 1910 Edison Studios came out with a film based on Mary Shelly's Frankenstein directed by J. Searle Dawley.

In 1918 came 'Alraune', a German silent science fiction horror film directed by Eugen Illés and Joseph Klein and starring Max Auzinger. A Professor specializing in 'genetics' conducts an experiment with a prostitute by impregnating her with a mandrake(the plant that is believed by legend to sprout from the semen of hanged prisoners.)Alarune is a beautiful lady but a sociopath born from this experiment and entices many men.

In 1918 came one of the first Danish sci fi movie Trip to Mars, however the next sci fi film came nearly 50 years afterwards in 1962 called Reptilicus.

The Danish film's advertisement even in those days in Motion Pictures News published on behalf of Tower Film Corp. New York, on July 19,1920, read , "A Trip to Mars – 'A Revelation', Truly a Remarkable Novelty Feature with an All Star Cast 5000 actors, 50 Gorgeous Settings, $ 100,000 worth of Mechanical Devices … You Can Now Make Reservations for Territorial Rights…." .

The story has Professor Planetarios going off to Mars on a spaceship which looks more like a dirigible and that too with a searchlight on top. He finds a race of white-robed, mystic vegetarians and converts to pacifism . He comes back to spread the message on earth. The film came soon after world had witnessed the horrors of the First World War.

In German science fiction films there came out a serial 'Homunculus' during the silent film era directed by Otto Rippert released during World War I with a plot similar to Frankenstein.

An American version of Jules Verne's novel '20,000 Leagues Under the Sea' also came out during the silent film era in 1916 directed by Stuart Paton.

Segundo de Chomó's directed 'Electric Hotel' was a French film released in 1908. The story in the silent era showed a couple Laure and Bertrand arrive at hotel and being pleasantly surprised to see that a control board allows inanimate objects to come to life. Several tasks are done automatically like polishing shoes, a brush styling hair and putting away of luggage. However the whole precision and system goes and chaos prevails due to a drunken hotel employee who messes up the switches.

In 1902 was released the French film Le Voyage dans la Lune (A Trip to the Moon) created by Georges Melies almost 70 years before Man landed on the moon, considered to be the first full fledged science fiction film. .

It shows a group of ambitious white bearded astronomers going to the moon in a space craft being launched to the moon with the help of a large cannon .The beginnings of scientific and industrial growth is shown in the film with scientists using telescopes and images of factories with smoking chimneys.

The waving row of cheering girls with buglers at the time the rocket is fired is reminiscent of an IPL event.

Interestingly they meet the moon people almost like the aliens in many of today's sci fi movies fight them before returning back to Earth to get a hero's welcome.

Despite it being a jerky silent movie the enthusiasm in the actors and the fast moving plot is still gripping as any sci-fi movie seen today.

About 8 years after this film USA came out with the five minute film 'A trip to Mars' made by Edison for his home Kanetoscope where an individual could view a film through a peephole. The film was directed by Ashley Miller.

The story is more of a fantasy rather than science fiction where a professor of chemistry is shown to discover that a combination of two powders leads to a product which reverses the well-known law of gravity, the container in which the powder is made itself rises up in the air. The scientist is very happy and tries out the experiment by putting the mixture on his chair and that too flies into the air.

The professor in his joy to announce his discovery and waves the two papers containing the powders over his head, a little of each is accidentally spilled on him. Before he can realize what is happening the professor is sailing out of the window and through the clouds, on his way to the planet Mars.

Arriving there upside-down he scrambles along the under surface of the planet and finally finds himself in a dense wood of trees, which are of enormous size. The upper halves of the trees are semi-human monsters which reach out long arms to seize him as he runs between their trunks. Escaping this danger he climbs upon a ledge of rocks which to his terror he finds is actually the face of a giant creature who opens his eyes and lips and blows the professor high in the air.

He catches the professor on his gigantic hand. The professor looks up at the enormous face above him and sinks on his knees.

Again the Martian blows his breath upon him, and slowly the professor congeals until he is a snowball, which the Martian places over the fire. The ball swells and swells until it explodes, and the professor is again blown off into space, but this time in the direction of Mother Earth.

We see him falling headlong and turning over and over in his homeward flight, until he drops into his laboratory again; and having had quite enough of reverse gravity, the professor casts the powders in rage upon the floor.

The result is that one side of the floor and the same side of the room rise until the professor slides down its perpendicular edge; and the last that is seen of him is the room is spinning him around at a dizzy pace.

In the later years of the 1950s, the major American studios limited themselves to adaptations of "classics" by Jules Verne and H. G. Wells , like 'The War of the Worlds' '20,000 Leagues Under the Sea' and 'Journey to the Center of the Earth'.

Science fiction literature also had a huge impact on early films. Books like Mary Shelly's Frankenstein(1910) and Robert Louis Stevenson's Dr. Jekyll and Mr Hyde (1913) were adapted into films, mixing sci-fi and horror together.

By 1930 you had movies with sound and production of short film serials like The Phantom Empire (1935), about a cowboy who stumbles upon a technologically advanced underground civilization with ray guns, robots and advanced TV's.

The science fiction films continued with similar themes of space travel, mad scientists and inventions and as is the trend today many successful films preferred to have many sequels to their films.

The bombing of Hiroshima and Nagasaki chillingly brought out the disastrous consequences associated with scientific advancements with imminent fallout on the science fiction films.

Effects of a nuclear war and dystopian future became a permanent theme of science fiction writers.

The 1950 film Destination Moon shows four Americans reaching moon on a rocket using nuclear power while competing with Russia. In the 1950s there were films with alien and UFOs were released.

In the early part of 1952 the first science fiction film in India Kaadu or The Jungle, came out as a Tamil-American co-production.

Directed by William Berke, Mr. T. R. Sundaram (The Modern Theatres Ltd) & William Berke Production. It stars Rod Cameron, Cesar Romero, Marie Windsor and M.N. Nambiar in lead roles. The film was the first science fiction film in India. With the adventurous story of a great white hunter and an Indian princess trekking into the Indian jungle to investigate deaths caused by wild animals. What they discover is pre historic woolly mammoths in the jungle who were causing the havoc.

In 1957 there came out in US science the fiction film 'The Incredible Shrinking Man' based on Richard Matheson's novel 'The Shrinking Man' directed by Jack Arnold. The film starred Grant Williams as Scott and Randy Stuart as Scott's wife Louise. Having come across radiation laden fog on the sea after a few months Scott starts shrinking to the extent that he becomes smaller than a cat and almost like an insect. The film became one of the highest grossing films of the decade.

In 1966 came out Fahrenheit 451 a science fiction film based on Ray Bradbury's novel which talked of a dystopian society in future where books are burned, and freedom of speech is restricted by government.

Another major sci fi film came out in 1966 called Fantastic Voyage (1966) which tells the story of the main character along with several others going inside the human body in a miniaturized submarine and exploring the inside of a human body to try and remove a blood clot in the brain to save the life of a scientist.

Planet of the Apes (1968) was also a very popular film that eventually resulted in four sequels and a TV series.

One of the most significant sci-fi films during 1960s is Stanley Kubrick's 2001: A Space Odyssey (1968). The film had vast canvas dealing with human evolution from the state of apes to almost becoming the star child travelling through the Universe.

It was groundbreaking for its realistic visual effects and portrayal of space travel.

Science fiction films continued to dwell with space travel following Man's landing on moon in 1969 and also several films started taking up contemporary subject matters of ecological threats and technological conflicts.

Steven Spielberg's E.T. the Extra Terrestrial (1982) was one of the most successful films in the 1980s.

Slowly the distinctions between science fiction films and fantasy and super hero films were being obliterated in the public mind with many super hero movies and films like Star Wars coming out.

In 1999 came The Matrix (1999), a blockbuster science fiction movie questioning the 'reality' of the outside world and positing a computer based virtual world.

Computers also started playing a significant part in film making process slowly taking over special effects and film editing and production.

In the last few decades of the twentieth century fantasy and superhero films become more popular though from time to time we also have some good science fiction films like Independence Day (1996), The Matrix(1999), The Matrix Reloaded (2003) and The Matrix Revolutions (2003), A.I. Artificial Intelligence (2001) and Minority Report (2002),Gravity (2013), Interstellar (2014), Inception (2010), among others.

Besides US many other countries all over the world have also produced science fiction films but it is beyond the scope of this book to go over them. Just to give some idea the robust French film industry has a renowned sci fi film, by director Chris Marker "La Jetée" ("The Jetty") which is extremely evocative and powerful made by using primarily still photos in black and white as a photo montage to explore the memory of a man in the aftermath of nuclear war .

The film has no dialogues and just one single shot in a movie camera rest of the movie has still shots. However the orchestra keeps the tension and the tempo of the movie alongwith the crisp editing with fade ins and fade outs. In 2010 the Time magazine ranked it among the top 10 time travel movies.

World renowned French film maker Jean-Luc Godard has also made a dystopian sci fi film, "Alphaville" . It is situated far away from earth where a U.S. secret agent is sent on a mission to find a missing person and also free Alphaville's citizens from repressive rule. The film won the Golden Bear award of the 15th Berlin International Film Festival in 1965.

Much before the very popular American film Hunger Games was released on a similar theme there was the French-Yugoslav film "Le Prix du Danger" ("The Prize of Peril") in 1983, directed by Yves Boisset based on Robert Sheckley's short story.

The shocking story was about men and women who volunteer to be hunted and be killed and if they survive they are to get huge prize. The whole show is part of live reality show.

Similarly another French film directed by Bertrand Tavernier "La Mort en Direct" (1980) also deals with reality show but in completely unique way. The hero has a camera implanted in his brain by a television studio so that anything he looks at is recorded. It becomes extremely poignant when he is asked to record the last days of a terminally ill patient just before her death.

However as the plot progresses the hero becomes blind and it is revealed that the terminally ill patient is not going to die but she commits suicide at the end.

The film is based on David G. Compton's novel, "The Unsleeping Eye."

If we look at the British science fiction films, in 1936 William Cameron Menzies, directed 'Things to Come' written by H.G.Wells, which was a kind of block buster of that period encompassing the destruction of the civilization in horrific wars and disease , then rebuilding of a new world and civilization with cities underground even though peace and scientific progress is there still there is opposition to progress as Man wants to go further to explore and conquer space.

The Day the Earth Caught Fire (1961) directed by Val Guest shows the destructive potential of nuclear testing as it is unearthed by an investigative journalist that the planet's axis has undergone a shift due to frequent nuclear detonations.

Japan is one country where science fiction, fantasy, horror seem to be part of its cinema culture. Innumerable monster and alien movies can be seen there merging the lines between the three genres of fantasy, horror and science fiction.

One of the cult film which is known throughout the world giving rise to large number of remakes and similar films is of course 1954 Godzilla, directed by Ishiro Honda, which is one of the longest running franchise with about 36 films, 32 made by Toho and four by American production houses . The film was extremely popular and influenced directors all over the world. The idea of a monstrous dinosaur waking up from the sea off the coast of Japan from hibernation due to atomic bomb tests caught the public imagination

The character of Godzilla has its own star on the Hollywood Walk of Fame.

In 1956 director Koji Shima released the film "Warning from Space" a sci fi film where aliens in their star fish shaped UFO circling Tokyo want to warn humankind regarding an impending strike by a meteor. An alien transforms itself as a lady singer and lands on Tokyo to warn people about the impending disaster so that they can avoid it.

In the same year Ishiro Honda released "Rodan" in which during a mining accident pre-historic giant insects and a giant reptile crawl out and attack Japan.

In 1957 Ishiro Honda's "The Mysterians" depicted how aliens come to help a village which is being destroyed by a rogue giant robot.

Director Strigeaki Hidaka and William Ross in 1960 released the film "World War III Breaks Out"showing a future scenario of US and communist block warring with each other and Japan getting drawn in.

Ishiro Honda came out with his film "The Human Vapour" where the hero almost in lines of the invisible man can turns into a gaseous form.

In 1961 film 'The Last War' again projects a future scenario of a war between Soviet Union and USA directed by Shue Matsabayashi.

A giant celestial body is hurtling towards earth is the theme of the movie 'Gorath' directed by Ishiro Honda, released in 1962. The movie has its tense moments as huge rocket thrusters are prepared to move earth from its orbit which is eventually successfully done. Interestingly even this movie of Honda has a pre historic giant walrus 'Maguma' coming out of its icy den.

In 1963 Ishiro Honda released King Kong versus Godzilla.

Science fiction Japanese film 'Face of Another' directed by Hiroshi Teshighara came out in 1966 depicting how changing the face results in personality change of the protagonist.

Australian film 'These Final Hours' 2013 written and directed by Zak Hilditch has completely different take on a disaster sci fi where it depicts the last few hours of people as the earth is hit by an asteroid leading to major fires which will engulf Perth, in Australia in next twelve hours.

An emotional and incisive take on what people would do in the last few hours of their life and the world. The hero even in the last few hours finds some meaning in his life which in a way helps him reconcile with his beloved and face his end calmly along with her.

The film raises issue of meaning of life and that of true love.

A Chinese science fiction film which has been very successful in recent years is 'The Wandering Earth' 2019 directed by Frant Gwo, based on a novel by Liu Cixin.

The film set in the future talks of a time when the sun is expanding and the earth is slowly moving out of its pull and would no longer be livable. The astronauts want to guide the Earth away from an expanding Sun, while attempting to prevent a collision with Jupiter. The film was released in China in 2019 and grossed 700 million dollars worldwide.

CHAPTER 5
DIRECTORS OF SCIENCE FICTION FILMS

Steven Spielberg

Science fiction films seem to have inspired several top directors in the world and many of them have tried to have at least one or two sci fi films in their directorial portfolio during their career. There are only few who have consistently made good sci fi films and riding as a colossus on the top is jewish producer, director, screen writer Steven Allan Spielberg.

To him goes the credit of directing the sci fi greats like ET (Extra Terrestrial), Close Encounters of the Third Kind, Jurrasic Park, Back to the Future apart from other blockbusters like Jaws (1975), Schindler's List (1993), Amistad (1997), Saving Private Ryan (1998),Munich (2005), Lincoln (2012), Indiana Jones Raiders of the Lost Ark (1981) and its later sequels.

Filming adventure films was a craze for Spielberg even as a schoolboy who used to make 8 mm adventure films during his school days. He made a home movie of a train crash with his own toy trains. Interestingly, in an interview later he even said that his desire for breaking things was fulfilled by destroying things in his films rather than in reality.

Born in Cincinnati, Ohio, in Dec. 18,1946, his father was an electrical engineer involved in computer development and mother ran a restaurant. He had a rough childhood suffering from bullying because of his jewish background.

Showing his talent early he made a nine minute 8 mm film in 1958 when he was a boy scout for the photography merit badge, titled _ "The Last Gunfight".

At the young age of just thirteen he won a prize for a 40-minute, war film he titled 'Escape to Nowhere'. No wonder his deep passion for film making was inherent and natural which continued throughout his life.

The maker of Jurassic Park has said that he was greatly influenced as a boy by the film Godzilla, King of the Monsters (1956), "the most masterful of all the dinosaur movies because it made you believe it was really happening", besides other adventure films like Captains Courageous , Pinocchio.

According to Steven Spiegel the movie that— "set me on my journey" was the large scale and adventurous epic historical movie Lawrence of Arabia (1962) produced by Sam Spiegel.

Spielberg wrote and directed his first independent film when he was just 16 years old and that too was a science fiction film _'Fireflight', a 140-minute film of UFOs , alien abductors who want to create human zoo in their planet Altaris. The movie was made at the expense of $ 500 funded by his father. The film was screened in the local theatre and earned $ 501. An astute businessman Spielberg's remarks about his first show — "I counted the receipts that night, we charged a dollar a ticket. Five hundred people came to the movie and I think somebody probably paid two dollars, because we made one dollar profit that night, and that was it." Ironically Spielberg who proudly counted that one dollar profit that memorable day went on to make millions and millions of dollars of profit from films on similar themes in his lifetime. His net worth is more than $3 billion today. Fireflight was clearly an inspiration for Close Encounters and other sci fi movies.

Spielberg studied at California State University Long Beach and got a job as intern with editing department in Universal Studios. At Universal Studios he got to write and direct a 26 minute film _ 'Amblin ' about a two hitchhiker couple's journey which was released in theatre and he got a number of awards.

trtrtrtrttrtr

Spielberg's first full-length film _The Sugarland Express', was not successful on the box office though praised by critics. An observant reviewer in The Hollywood Reporter who could spot talent had even then remarked that "a major new director is on the horizon."

Richard D. Zanuck and David Brown offered Spielberg to be the director of Jaws, a thriller-horror film based on the Peter Benchley novel about an enormous killer shark.

That single opportunity was the turning point of Spielbereg's career as it turned out to be a huge success and won three Academy Awards for editing, original score and sound. It raked in $470 million worldwide at the box office. Instantly Spielberg became a millionaire and was looked upon as a top director. Speilberg wrote and directed _"Close Encounters of Third Kind" in 1977. This was again a big hit dealing with the subject of extra terrestrials. Critics appreciated the film and Spielberg established himself as a director.

Spielberg got his first Best Director nomination from the Academy as well as earning six other Academy Awards nominations. It finally won

Oscars in two categories (Cinematography, Vilmos Zsigmond, and a Special Achievement Award for Sound Effects Editing, Frank E. Warner).

In 1980s Spielberg teamed with Star Wars creator and friend George Lucas on an action adventure film, Raiders of the Lost Ark, the first of the Indiana Jones films. It became the biggest film at the box office in 1981 and was nominated for Best Director ,Best Picture at Oscars.

Spielberg's next big science fiction film was _"Extra Terrestrial"' which won him Best Director Award plus the film got three other Academy Awards including Best Picture award. Extra Terrestrial was extremely successful in the box office and become one of the top grossing films of all time.

Speilberg's success stories continued with several blockbuster films like Poltergeist, The Twilight Zone, Indiana Jones and the Temple of Doom.

The fact that Spielberg could take up serious stories was established with The Colour Purple, an adaptation of Alice Walker's Pulitzer Prize-winning novel.

His film _ "Indiana Jones and the Last Crusade " was a huge international success on the box office getting more than even Batman worldwide.

Spielberg's first romantic film was Always in 1989, with actor Richard Dreyfuss. In 1991, Spielberg's fantasy film Hook was hugely successful and made $ 300 million in profits.

But soon after came his science fiction film _ "Jurrasic Park". In 1993 he turned Michael Crichton's novel into one the world's biggest box office hits earning $ 914.7 million worldwide. The genetically engineered dianosaur theme awed audiences all over the globe and led to a large number of films following similar themes.

One of the serious films made by Spielberg _"Schindler's List' was based on the true story of Oskar Schindler, a man who risked his life to save 1,100 Jews from the Holocaust. It earned Spielberg his first Academy Award for Best Director and it also won the Best Picture prize.

Spielberg used the profits to set up the Shoah Foundation, a non-profit organization that archives filmed testimony of Holocaust survivors.

In 1997, the American Film Institute listed it among the 10 Greatest American Films ever made . In 1996, Spielberg directed the sequel to 1993's Jurassic Park with "The Lost World: Jurassic Park",' which generated over $618 million worldwide despite mixed reviews, and was the second biggest film of 1997 behind James Cameron's Titanic.

He took renowned actor Tom Hanks in _"Saving Private Ryan', a world war II film which was highly acclaimed by critics and became the second highest grossing film in 1998 worldwide with $481.8 million because of its intense and realistic portrayal of war.

This film won Spielberg his second Academy Award for his direction. Several other films like Black Hawk Down and Enemy at the Gates are supposed to have been influenced by its graphic depiction of combat scenes.

In 2001, Spielberg filmed fellow director and friend Stanley Kubrick's final project, A.I. _"Artificial Intelligence" which Kubrick was unable to begin during his lifetime.

Throughout the 1980s and early 1990s, Kubrick had collaborated with Brian Aldiss on an expansion of his short story "Supertoys Last All Summer Long" about a robot that resembles and behaves as a child, and his efforts to become a 'real boy' in a manner similar to Pinocchio.

Kubrick had approached Spielberg in 1995 with the AI script with the possibility of Steven Spielberg directing it and Kubrick producing it as the subject matter was closer to Spielberg's sensibilities than his. Following Kubrick's death in 1999, Spielberg took the various drafts and notes left by Kubrick and his writers and composed a new screenplay.

He is supposed to have said that Kubrick seemed to be guiding him "I felt like I was being coached by a ghost." A huge success the futuristic film about a humanoid android longing for love, internationally grossed $236 million.

In the next film Spielberg took actor Tom Cruise for the futuristic science fiction film _"Minority Report'" that earned over $358 million worldwide.

Another renowned actor Leonardo DiCaprio was chosen by Spielberg for his film _"Catch Me If You Can"'. Spielberg took Tom Hanks along with Catherine Zeta-Jones and Stanley Tucci in _"The Terminal",'.

In 2005, Spielberg directed a modern adaptation of "War of the Worlds" based on the H. G. Wells book of the same name starring Tom Cruise and Dakota Fanning, it made over $591 million worldwide.

Spielberg's film "Munich", about the massacre of Jewish players following the 1972 Olympic Games, received five Academy Awards nominations.

Spielberg then directed "Indiana Jones and the Kingdom of the Crystal Skull" earning $786 million worldwide.

Historical film Lincoln of Spielberg won the award for Best Production Design and Day-Lewis won the Academy Award for Best Actor for his portrayal of Lincoln.

In 1996, Spielberg shot original footage for a movie-making simulation game called Steven Spielberg's Director's Chair.

Spielberg made films with a broad canvas and strong emotive feeling ,being able to convey a childlike sense of wonder like Close Encounters of the Third Kind, E.T. the Extra-Terrestrial, Hook, A.I. Artificial Intelligence and The BFG. Tension between father and son relationship was also a theme in his films reflecting back to his parents divorce.

In 2001, he was appointed as an honorary Knight Commander of the Order of the British Empire (KBE) by Queen Elizabeth II for services to the entertainment industry of the United Kingdom.

In 2004, he was admitted as knight of the Légion d'honneur by French president Jacques Chirac. On July 15, 2006, Spielberg was also awarded the Gold Hugo Lifetime Achievement Award at the Summer Gala of the Chicago International Film Festival. In 2005, Steven Spielberg was rated the greatest film director of all time by Empire magazine.

James Cameron

Canadian filmmaker, James Francis Cameron who left his job as a truck driver to enter the film industry after watching Star Wars in 1977, is best known for making landmark films like Avatar,The Terminator and Titanic.

Avatar and Titanic were for a long time the second and third highest-grossing films of all time, earning $2.78 billion and $2.19 billion, respectively.

Son of an electrical engineer Philip Cameron of Scottish descent, James Cameron was born on August 16, 1954 in Kapuskasing, Ontario. His mother Shirley was an artist and nurse.

He left Fullerton College which he joined to study Physics within a year even after switching to study English and took up jobs of a truck driver and a janitor. Passionate for films he studied film technology and special effects during his free time in library.

After borrowing money from a consortium of dentists, he wrote, directed and produced his first short film, Xenogenesis (1978) with a friend. Cameron started in the film line when he got a job as a miniature model maker at Roger Corman Studios. He was soon employed as an art director for the science-fiction film _"Battle Beyond the Stars' (1980).

He carried out the special effects for John Carpenter's Escape from New York (1981), served as production designer for Galaxy of Terror (1981), and was consulted on the design for Android (1982).

He became director of a major film when he was made special effects director for the sequel to Piranha (1978), titled "Piranha II: The Spawning in 1982." Shot in Rome, Italy and on Grand Cayman Island.

It is really strange to know that the idea of Terminator came to him in his dreams actually a nightmare. He had a nightmare about an invincible robot hit-man sent from the future to assassinate him. This event inspired him to write the script of The Terminator in 1982.

His wife Gale Anne Hurd, a colleague and founder of Pacific Western Productions, produced the film The Terminator . It was a box office success and earned over $78 million worldwide In 2008, the film was selected for preservation in the United States National Film Registry, being deemed "culturally, historically, or aesthetically significant".

In 1984, Cameron helped in writing the screenplay of Rambo: First Blood Part II with Sylvester Stallone.

Cameron then directed the sequel to Alien (1979), a science fiction horror by Ridley Scott. After titling the sequel Aliens (1986), Cameron recast Sigourney Weaver as Ellen Ripley, who first appeared in Alien.

Aliens was a box office success, generating over $130 million worldwide. The film was nominated for seven Academy Awards in 1987; Best Actress, Best Art Direction, Best Film Editing, Best Original Score and Best Sound. It won awards for Best Sound Editing and Best Visual Effects .

After Aliens, Cameron directed 'The Abyss', intelligent life form under the ocean. Abyss was praised for its special effects, and earned $90 million worldwide. It received four Academy Award nominations and won Best Visual Effects.

Cameron also directed and produced Terminator 2: Judgment Day (1991) which was written by Cameron and William Wisher Jr.

Terminator 2 was one of the most expensive films to be produced, costing at least $94 million. Terminator 2 broke box office records earning over $300 million worldwide. It won four Academy Awards: Best Makeup, Best Sound Mixing, Best Sound Editing, and Best Visual Effects.

In 1994, Cameron and Schwarzenegger reunited for their third collaboration, True Lies, the film earned $232 million worldwide. The film was nominated for an Academy Award for Best Visual Effects and Jamie Lee Curtis won a Golden Globe Award for Best Actress.

His next major project was Titanic (1997), an epic film about RMS Titanic which sank in 1912 after striking an iceberg. With a production budget of $200 million, Titanic is one of the most expensive films ever made.

Starting in 1995, Cameron took several dives to the bottom of the Atlantic Ocean to capture footage of the wreck, which would later be used in the film.

The film received strong critical acclaim and earning over 2 billion dollars became the highest-grossing film of all time, and held this position for 12 years until Cameron's Avatar beat the record in 2010.

The costumes and sets were praised, and The Washington Post considered the CGI graphics to be spectacular. It won 11Academy awards (tying the record for most wins with Ben-Hur (1959) and later, The Lord of the Rings: The Return of the King (2003), including: Best Picture, Best Director, Best Art Direction, Best Cinematography, Best Visual Effects, Best Film Editing, Best Costume Design, Best Sound Mixing, Best Sound Editing, Best Original Score, and Best Original Song.

Cameron has made many documentaries on the deep sea, one of Cameron's personal interests.

His next film was Avatar, with the story line set in the mid-22nd century, with a budget of around $300 million. The cast includes Sam Worthington, Zoe Saldana, Stephen Lang, Michelle Rodriguez and Sigourney Weaver. It was composed with a mix of live-action footage and computer-generated animation.

Released in 2009 it broke box office records and earned more than $2.74 billion worldwide, becoming the highest-grossing film of all time surpassing Titanic. It was the first film to ever earn more than $2 billion worldwide. Avatar was nominated for nine Academy Awards, including Best Picture and Best Director, and won three: Best Art Direction, Best Cinematography, and Best Visual Effects.

Paying his gratitude to the science fiction genre he has once stated — "|Without Jules Verne and H. G. Wells, there wouldn't have been Ray Bradbury or Robert A. Heinlein, and without them, there wouldn't be George Lucas, Steven Spielberg, Ridley Scott or me".

Cameron is an expert on deep-sea exploration and has contributed to advancements in underwater filming and remotely operated vehicles and helped develop the 3D Fusion Camera System.

In 2011, Cameron became a National Geographic explorer-in-residence.

Cameron's films are often based on themes which explore the conflicts between intelligent machines and humanity or nature, dangers of corporate greed, strong female characters, and a romance subplot. Dramatic moments in wide open sea is explored in The Abyss and Titanic.

A species of frog, Pristimantis james cameroni, was named after Cameron for his work in promoting environmental awareness and advocacy of veganism.

In 2013, Cameron received the Nierenberg Prize for Science in the Public, which is annually awarded by the Scripps Institution of Oceanography. In 2019, Cameron was appointed as a Companion of the Order of Canada by Governor General Julie Payette.

Alfonso Cuarón

Mexican filmmaker and screenwriter Alfonso Curaon has made a name for himself in the world of science fiction film makers by directing two science fiction films known as Gravity and Children of Men.

He had both his parents from the field of science with his mother Cristina Orozco being a pharmaceutical biochemist and father Alfredo Cuarón being a doctor specializing in nuclear medicine.

Interestingly there was also a mix of science and film among his siblings with one of the brothers being a filmaker , Carlos, and another brother Alfredo being a conservation biologist.

Cuarón studied philosophy at the National Autonomous University of Mexico (UNAM) and filmmaking at CUEC (Centro Universitario de Estudios Cinematográficos).

After a period of making short films and working for television, in 1995, Cuarón worked on two films first one based on an adaptation of Frances Hodgson Burnett's classic novel A Little Princess, and the second an adaptation, of Charles Dickens's Great Expectations starring Ethan Hawke, Gwyneth Paltrow and Robert De Niro.

In 2004, Cuarón directed the third film in the successful Harry Potter series, Harry Potter and the Prisoner of Azkaban. Cuaron took P.D.James novel 'Children of Men', and adapted it into a movie with star cast consisting of Clive Owen, Julianne Moore and Michael Caine depicting a dystopian (very likely looking at some of the present day fertility trends), where all of humanity has gone infertile. The film was a great success receiving three Academy Award nominations.

The next film of Cuaron could possibly be one of the most spectacular and realistic movies made on space Gravity. Several of his films have received critical acclaim and accolades. He has been nominated for 11 Academy Awards winning four of them, including two Best Director awards for Gravity (2013) and Roma (2018).

He is the first Latin American director to receive the award for Best Director. He has also received Academy Awards for Best Film Editing for Gravity and Best Cinematography for Roma. Cuarón has been nominated for Academy Awards in six different categories, a record he shares with Walt Disney and George Clooney.

Cuaron actually brought in his own style of filming which makes the realism in his films striking. In the era of fast paced cuts in between shots, he prefers lengthy shots sometimes continuing for several minutes. There is an unhurried languid movement of the camera as it takes the viewer from long shots to extreme close up, then some wandering shots placing the character and action in a specific location and environment where we can glimpse other people or action taking place.

In May 2015, Cuarón was announced as the President of the Jury for the 72nd Venice International Film Festival.

Cuaron's film Roma debuted at 75th Venice International Film Festival and was highly successful it got two Golden Globes (Best Foreign Language Film and Best Director for Cuarón) and three Academy Awards (Best Director, Best Foreign Language Film, and Best Cinematography for Cuarón) out of a leading ten nominations.

Wachowskis

Lana Wachowski and Lilly Wachowski directed their magnum opus The Matrix in 1999. The sisters Lana Wachowski and and Lilly Wachowski both sisters are trans woman and are American film and television directors, writers and producers.

The Wachowskis made their directing debut in 1996 with Bound, and achieved fame with their second film The Matrix (1999), a major box office success for which they won the Saturn Award for Best Director.

They wrote and directed its two sequels: The Matrix Reloaded and The Matrix Revolutions (both in 2003), and were involved in the writing and production of other works in that franchise.

Following the commercial success of The Matrix series, they wrote and produced the 2005 film V for Vendetta (an adaptation of the graphic novel by Alan Moore & David Lloyd), and in 2008 released the film Speed Racer, a live-action adaptation of the Japanese anime series. Their next film, Cloud Atlas, based on the novel by David Mitchell and co-written and co-directed by Tom Tykwer, was released in 2012.

Their mother, Lynne was a nurse and painter. Their father, Ron Wachowski, was a businessman of Polish descent. The Wachowskis attended the Kellogg Elementary School in Chicago's Beverly area, and graduated from Whitney Young High School, known for its performing arts and science curriculum, in 1983 and 1985, respectively. Lana went to Bard College in New York and Lilly attended Emerson College in Boston.

They completed The Matrix, a science fiction action film, in 1999. The movie stars Keanu Reeves as Neo, a hacker recruited by a rebellion to aid them in the fight against machines who have taken over the world and placed humanity inside a simulated reality called "the Matrix". Laurence Fishburne, Carrie-Anne Moss, Hugo Weaving and Joe Pantoliano also star in The Matrix.

The movie was a critical and commercial hit for Warner Bros. It won four Academy Awards, including for "Best Visual Effects" for popularizing the bullet time visual effect. It also got awards for Best Sound Mixing, Best Film Editing,Best Sound Editing. The Matrix came to be a major influence for action movies and has appeared in several "greatest science fiction films" lists.

In 2012, the film was selected for preservation in the National Film Registry by the Library of Congress for being "culturally, historically, and aesthetically significant."

After its success, the Wachowskis directed two sequels back-to-back, The Matrix Reloaded and The Matrix Revolutions.

Christopher Nolan

Despite being much younger to Steven Spielberg Christopher Nolan is perhaps the other top sci fi director who falls in his category though Spielberg may surpass Nolan with his broader canvas of films. Nolan's Inception is supposed to have broken new ground by making a riveting thriller with dream within a dream theme.

Christopher Edward Nolan CBE was born 30 July 1970is a British-American film director, producer and screenwriter known for making personal, distinctive films within the Hollywood mainstream. His directorial efforts have grossed more than US$5 billion worldwide, garnered 34 Oscar nominations and ten wins.

Born and raised in London, Nolan developed an interest in filmmaking from a young age. After studying English literature at University College London, he made his feature debut with Following (1998).

Nolan gained international recognition with his second film, Memento (2000), for which he was nominated for the Academy Award for Best Original Screenplay. His next film Insomnia (2002) a psychological thriller raked in $ 113 million worldwide against a production budget of $ 46 million and was critically acclaimed particularly the performances of Robin Williams and Al Pacino.He found further critical and commercial success with The Dark Knight Trilogy (2005–2012), The Prestige (2006), and Inception (2010), which received eight Oscar nominations, including for Best Picture and Best Original Screenplay. This was followed by Interstellar (2014), Dunkirk

(2017), and Tenet (2020). He earned Academy Award nominations for Best Picture and Best Director for his work on Dunkirk.

He has co-written several of his films with his brother Jonathan, and runs the production company Syncopy Inc. with his wife Emma Thomas. Time magazine named him as one of the 100 most influential people in the world in 2015, and in 2019, he was appointed Commander of the Order of the British Empire for his services to film.

He began making films at age seven, borrowing his father's Super 8 camera and shooting short films. From the age of eleven, he aspired to be a professional filmmaker inn his teenage years, Nolan started making films with Adrien and Roko Belic. Nolan and Roko co–directed the surreal 8 mm film Tarantella (1989), which was shown on Image Union, an independent film and video showcase on the Public Broadcasting Service.

He screened 35 mm feature films during the school year and used the money earned to produce 16 mm films over the summers.

Nolan studied English literature at University College London (UCL).

After earning his bachelor's degree in English literature in 1993, Nolan worked as a script reader, camera operator, and director of corporate videos and industrial films.

In 1995, he began work on the short film Larceny, which appeared at the Cambridge Film Festival in 1996 and is considered one of UCL's best shorts.

Memento (2000), was his major film that premiered at the Venice International Film Festival with its stars Guy Pearce and Carrie-Anne Moss in September 2000 and was greatly appreciated and was a box-office success. It received Academy Award and Golden Globe Award nominations for its screenplay, Independent Spirit Awards for Best Director and Best Screenplay.

In 2017, the film was selected by the Library of Congress for preservation in the United States National Film Registry, being deemed "culturally, historically, or aesthetically significant".

Impressed by his work on Memento, Steven Soderbergh recruited Nolan to direct the psychological thriller Insomnia (2002), starring Academy Award winners Al Pacino, Robin Williams, and Hilary Swank.

In 2003 Warner Bros. accepted Nolan's idea for a film on Batman with Christian Bale,Michael Caine, Gary Oldman, Morgan Freeman, and Liam Neeson, and came out with critically and commercially successful film Batman Begins. It was that year's ninth-highest-grossing film worldwide.

It was nominated for the Academy Award for Best Cinematography, and has been cited as one of the most influential films of the 2000s.

Before returning to the Batman franchise for a sequel, Nolan directed, co-wrote, and produced suspense thriller The Prestige (2006), an adaptation of the Christopher Priest novel about two rival 19th-century magicians. Starring Hugh Jackman and Christian Bale in the lead roles, The Prestige received critical acclaim and earned over $109 million worldwide.

In 2006, Nolan came up with sequel of Batman Begins called The Dark Knight which earned over $1 billion worldwide.

At the 81st Academy Awards, the film was nominated for eight Oscars, winning two: the Academy Award for Best Sound Editing and a posthumous Academy Award for Best Supporting Actor for Heath Ledger.

The Warner Bros. signed Nolan to direct Inception (2010) which with a star like Leonardo DiCaprio, became a commercial and critical success with a critic calling it "one of the best movies of the 21st century.

It made over $830 million worldwide and was nominated for eight Academy Awards, including Best Picture and Best Original Screenplay; it won the award for Best Cinematography, Best Sound Mixing, Best Sound Editing and Best Visual Effects. Nolan was also nominated for BAFTA and Golden Globe awards.

In 2012, Nolan directed his third and final Batman film, The Dark Knight Rises, with Christian Bale in the title role. The film grossed more than a billion dollars. Then Nolan produced Man of Steel (Superman) for Warner Bros with Zack Snyder to direct starring Henry Cavill, Amy Adams, Kevin Costner, Russell Crowe, and Michael Shannon, the film fetched more than $660 million at the worldwide.

Nolan next directed, wrote, and produced the science-fiction film Interstellar (2014) dealing with wormholes and blackholes in consultation with theoretical physicist Kip Thorne. Starring Matthew McConaughey, Anne Hathaway, Jessica Chastain, Bill Irwin, Michael Caine,and Ellen Burstyn, the film grossed over $690 million worldwide.

At the 87th Academy Awards, the film won Best Visual Effects and received four other nominations – Best Original Score, Best Sound Mixing, Best Sound Editing and Best Production Design.

Christopher Nolan is also passionate about film as an art form and preservation and distribution of the films of lesser-known filmmakers.

Dunkirk directed by Christopher Nolan was released in 2017 about World War II evacuation of Allied soldiers from the beaches of Dunkirk, France, in 1940. It became the highest grossing World War II film ever.

In 2020 came the blockbuster film Tenet whopping budget of $ 200 million the science fiction action thriller was produced by Christopher Nolan along with Emma Thomas , written and directed by Christopher Nolan. The film is about a secret agent who learns to manipulate the flow of time and prevents future annhilation of the world. The film grossed $ 363.7 million worldwide.

Jon Favreau

Today one of the top directors is another Jewish director Jon Favreau in the field of sci fi with movies like Iron Man, Daredevil, Spiderman, Avengers, Iron Man 2, Cowboys and Aliens and Disney series The Mandalorian.

Born in New York he was the son of Madeleine, an elementary school teacher and Charles Favreau, a special education teacher. His mother was Jewish from Russian descent and his father is a Catholic of Italian and French-Canadian ancestry. Favreau dropped out of Hebrew school to pursue acting and attended Queens College from 1984 to 1987 and in the summer of 1988, moved to Chicago to pursue a career in comedy where he got his first film role in Rudy (1993). He moved to Los Angeles and became actor-screenwriter with the film Swingers.

In 1997, he appeared on the television sitcom Friends, portraying Pete Becker. Favreau did the role of Gus Partenza in Deep Impact (1998). In 1999, he starred in the television film Rocky Marciano, based on the life of world heavyweight champion, Rocky Marciano. He later appeared in Love & Sex (2000), co-starring Famke Janssen.

Favreau appeared in 2000's The Replacements as maniacal linebacker Daniel Bateman.

He also starred in The Big Empty (2003), directed by Steve Anderson. Favreau is credited as a screenwriter for the film The First $20 Million Is Always the Hardest that premiered in 2002.

In the fall of 2003, he scored his first financial success as a director of the hit comedy Elf starring Will Ferrell, Zooey Deschanel, James Caan, and Peter Dinklage.

In April 2006, Favreau was signed to direct Iron Man movie which was released on May 2, 2008, the film was a huge critical and commercial success.

Iron Man was the first Marvel-produced movie under their alliance with Paramount, and Favreau served as the director and an executive producer. During early scenes in Iron Man, Favreau appears as Tony Stark's driver, Happy Hogan and also directed and executive produced the film's sequel, Iron Man 2. Favreau said in December 2010 that he would not direct Iron Man 3 but remain an executive producer.

Favreau was the third director attached to John Carter, in the film adaptation of Edgar Rice Burroughs' novel, and directed in 2009 Cowboys &Aliens,starring Daniel Craig and Harrison Ford.

In 2016, Favreau directed and produced The Jungle Book, for Walt Disney Pictures, which was released on April 15, 2016, which was a huge success.

He returned as Happy Hogan in the film Spider-Man: Homecoming (2017). He also was the co-executive producer for Avengers: Infinity War (2018). He was the executive producer for Avengers : Endgame in 2019, where he also had a role as Happy Hogan.

He was the executive producer and writer of Star Wars television series, titled The Mandalorian.

Denis Villeneuve

Canadian director Denis Villeneuve was born in village Gentilly in Bécancour, Quebec, to a notarian Jean Villeneuve and Nicole born on October 3, 1967.

Interested in sci-fi comics by French writers in childhood he soon developed a passion for cinema and made short films during high school. His favourite films included 2001: A Space Odyssey, Close Encounters of the Third Kind and Blade Runner. He studied cinema at the Quebec University at Montreal.

Villeneuve began his career making short films and won Radio-Canada's youth film competition, La Course Europe-Asie, in 1991.

He is a four-time recipient of the Canadian Screen Award for Best Direction, for Maelström in 2001, for Polytechnique in 2009, for Incendies in 2011 and for Enemy in 2013.

The first film directed by him _ "August 32nd on Earth" in 1998, premiered in the Un Certain Regard section at the 1998 Cannes Film Festival and Alexis Martin won the Prix Jutra for Best Actor. The film was selected as the Canadian entry for the Best Foreign Language Film at the 71st Academy Awards, but was not nominated.

His second film, Maelström (2000), attracted further attention and screened at festivals worldwide, ultimately winning eight Jutra Awards and the award for Best Canadian Film from the Toronto International Film Festival. The Jutra Awards are given to films in the Quebec film industry. His next film was Polytechnique about the shootings that occurred at the Montreal university. The film premiered at the Cannes Film Festival and received numerous honours, including nine Genie Awards, becoming Villeneuve's first film to win the Genie award for Best Motion Picture.

The film went on to win eight awards at the 31st Genie Awards, including Best Motion Picture, Best Direction, Best Actress (LubnaAzabal), Best Adapted Screenplay, Cinematography, Editing, Overall Sound, and Sound Editing.

Villeneuve directed Incendies a war film in 2010 and followed that with directing the crime thriller film Prisoners, starring Hugh Jackman and Jake Gyllenhaal. The film screened at festivals across the globe, won several awards, and was nominated for the Academy Award for Best Cinematography in 2014.

Villeneuve won Best Director for his sixth film, the psychological thriller Enemy (2014), at the 2nd Canadian Screen Awards. The film was awarded the $100,000 cash prize for best Canadian film of the year by the Toronto Film Critics Association in 2015.

Villeneuve subsequently directed his eighth film, Arrival (2016), based on the short story Story of Your Life by author Ted Chiang, from an adapted script by Eric Heisserer, with Amy Adams and Jeremy Renner starring. Arrival grossed $203 million worldwide and received critical acclaim, specifically for Adams's performance, Villeneuve's direction, and the film's exploration of communicating with extraterrestrial intelligence. Arrival appeared on numerous critics' best films of the year lists, and was selected by the American Film Institute as one of ten films of the year. It received eight nominations at the 89th Academy Awards, including Best Picture, Best Director, and Best Adapted Screenplay, ultimately winning one award for Best Sound Editing.

It was also awarded the Ray Bradbury Award for Outstanding Dramatic Presentation and the Hugo Award for Best Dramatic Presentation in 2017. In February 2015, it was announced that Villeneuve would direct Blade Runner 2049, the sequel to Ridley Scott's Blade Runner (1982).

For his work on Arrival, he received an Academy Award nomination for Best Director. He was awarded the prize of Filmmaker of the Decade by the Hollywood Critics Association in December 2019.

Andrei Arsenyevich Tarkovsky

Tarkovsky is considered one of the greatest film director and writer of the world cinema. Born 4 April 1932 his films are known for taking up significant spiritual and metaphysical themes. His films are known for their slow pace, dreamlike visual images, very long takes.

Tarkovsky studied film at Moscow's State Institute of Cinematography under filmmaker Mikhail Romm, and subsequently directed his first five feature films in the Soviet Union: Ivan's Childhood (1962), Andrei Rublev (1966), Solaris (1972), Mirror (1975), and Stalker (1979).

After years of creative conflict with state film authorities, Tarkovsky left the country in 1979 and made his final two films abroad; Nostalghia (1983) and The Sacrifice (1986) were produced in Italy and Sweden respectively. In 1986, he also published a book about cinema and art entitled Sculpting in Time. He died of cancer later that year.

Tarkovsky was the recipient of several awards at the Cannes Film Festival throughout his career and winner of the Golden Lion award at the Venice Film Festival for his first film Ivan's Childhood. In 1990, he was posthumously awarded the Soviet Union's prestigious Lenin Prize.

Three of his films—Andrei Rublev, Mirror, and Stalker—featured in Sight & Sound's 2012 poll of the 100 greatest films of all time.

Andrei Tarkovsky was born in the village of Zavrazhye in the Yuryevetsky District, Russia to poet Arseny Alexandrovich Tarkovsky, and Maria Ivanova Vishnyakova, a graduate of the Maxim Gorky Literature Institute. Tarkovsky studied Arabic at the Oriental Institute in Moscow.In 1954, Tarkovsky applied at the State Institute of Cinematography (VGIK) and was admitted to the film-directing program. Tarkovsky got an opportunity to see films of the Italian neo realists, French New Wave and of directors such as Kurosawa , Burnel, Bergman, Bresson, Andrezej Wajda, and Mizoguchi. He was in the same class as Irma Raush whom he married in April 1957.

In 1956 Tarkovsky directed his first student short film, The Killers, from a short story of Ernest Hemingway. The short film There Will Be No Leave Today and the screenplay Concentrate followed in 1958 and 1959.

Tarkovsky's first feature film was Ivan's Childhood in 1962. The film earned Tarkovsky international acclaim and won the Golden Lion award at the Venice Film Festival in the year 1962.

In 1965, he directed the film Andrei Rublev about the life of Andrei Rublev, the fifteenth-century Russian iconic painter. In 1972, he completed Solaris, an adaptation of the novel Solaris by StanisławLem. The film won the Grand Prix Spécial du Jury at the Cannes Film Festival and the FIPRESCI prize, and was nominated for the Palme d'Or. The last film Tarkovsky completed in the Soviet Union was Stalker.

Tarkovsky's film Nostalghia was presented at the Cannes Film Festival in 1983 and won the FIPRESCI prize and the Prize of the Ecumenical Jury. Tarkovsky also shared a special prize called Grand Prix du cinéma de creation with Robert Bresson.

The Sacrifice was Tarkovsky's last film dedicated to his son, Andrei Jr. released after his death and it received at the Cannes Film Festival the Grand Prix Spécial du Jury, the FIPRESCI prize and the Prize of the Ecumenical Jury.

Renowned director Ingmar Bergman, a renowned director has commented on him, "Tarkovsky is for me the greatest, the one who invented a new language, true to the nature of film, as it captures life as a reflection, **life as a dream."**

Stanely Kubrick

Director Stanely Kubrick has just done one pure science fiction film but that single movie virtually immortalized his name among the makers of the greatest sci fi films ever. The name of the film is — "2001 : A Space Odyssey.""

Kubrick was born in New York City to a Jewish family. He was the first of two children of Jacob Leonard Kubrick of Polish and Austrian Jewish descent his wife Sadie Gertrude Kubrick.

A keen photographer from young days he was able to transform short stories and novels into visually stunning movies which ranked him as one of the greatest filmmakers in cinematic history.

He grew up in the Bronx, New York City, and attended William Howard Taft High School from 1941 to 1945. Introverted and shy, Kubrick had a low attendance record and often skipped school to watch double-feature films. His father was a homeopathic doctor. He displayed an interest in literature from a young age and began reading Greek and Roman myths and the fables.

An average student he was interested in literature, photography, and film. At the age of 13, Kubrick's father bought him a Graflex camera, triggering a fascination with still photography. While still in high school Kubrick was chosen as an official school photographer Eventually he sold a photographic series to Look magazine, having brought a photo to Helen O'Brian, the head of the photographic department, who then purchased it for £25. In 1946, he became an apprentice photographer.

He was inspired by the complex, fluid camerawork of the director Max Ophüls, whose films influenced Kubrick's later visual style, and by the director Elia Kazan, whom he described as America's "best director" at that time, with his ability of "performing miracles" with his actors.

Kubrick had become obsessed with the art of filmmaking. Sergei Eisenstein's theoretical writings had a profound impact on Kubrick, and he took a great number of notes from books in the library of Arthur Rothstein, the photographic technical director of Look magazine. He learnt all aspects of film production and direction after passing high school.

After working as a photographer for Look magazine in the late 1940s and early 1950s, he began making short films on a shoestring budget, and made his first major Hollywood film, The Killing, for United Artists in 1956. This was followed by two collaborations with Kirk Douglas—the war picture Paths of Glory (1957) and the historical epic Spartacus (1960).

With a creative mind of his own he soon had differences with Douglas and the film studios and moved to the United Kingdom in 1961. His first productions in Britain were two films with Peter Sellers, Lolita (1962) and Dr. Strangelove (1964).

A perfectionist, Kubrick assumed control over most aspects of the filmmaking process, from direction and writing to editing, and took painstaking care with researching his films and staging scenes, working in close coordination with his actors and other collaborators. He often asked for several dozen retakes of the same shot in a movie, which resulted in many conflicts with his casts.

Despite the resulting notoriety among actors, many of Kubrick's films broke new ground in cinematography. The scientific realism and innovative special effects of 2001: A Space Odyssey (1968) were unprecedented in the history of cinema, and the film earned him his only personal Oscar, for Best Visual Effects. Steven Spielberg has referred to the film as his generation's "big bang"; it is regarded as one of the greatest films ever made.

For the 18th-century period film Barry Lyndon (1975), Kubrick obtained lenses developed by Zeiss for NASA, to film scenes under natural candlelight.

With The Shining (1980) which he directed and produced he became one of the first directors to make use of a Steadicam for stabilized and fluid tracking shots. While many of Kubrick's films were controversial including A Clockwork Orange (1971) which had to be pulled out from circulation in the UK most were nominated for Oscars, Golden Globes, or BAFTA Awards, and underwent critical reevaluations.

Kubrick's Dr. Strangelove or: How I Learned to Stop Worrying and Love the Bomb (1964) has been named as the sixth-best comedy film of all time by The Guardian in 2010.

Kubrick did not elaborate on the meaning of his films and left it for variety of interpretation by the viewers and critics. He knew the importance of silence to drive home a point than dialogue and therefore there were many moments in his films where the emphasis was on images and sound rather than any verbalization. His films were universal because they touched the emotional chordin every human being. Kubrick being a perfectionist would go for multiple takes for his shots till he was satisfied sometimes as many as fifty takes for one scene.

Kubrick is also known for using "one-point perspective", which leads the viewer's eye towards a central vanishing point. The technique relies on creating a complex visual symmetry using parallel lines in a scene which all converge on that single point, leading away from the viewer. Combined with camera motion it could produce an effect that one writer describes as "hypnotic and thrilling". This point of view which is naturally riveting keeps the audience deeply engrossed.

The Shining was among the first few movies use the then-revolutionary Steadicam . Kubrick used it to its fullest potential, which gave the audience smooth, stabilized, motion-tracking by the camera. Kubrick described Steadicam as being like a "magic carpet", allowing "fast, flowing, camera movements" in the maze in The Shining which otherwise would have been impossible.

Kubrick is credited with introducing Hungarian composer György Ligeti to a broad Western audience by including his music in 2001, The Shining and Eyes Wide Shut.

Kubrick died on March 7, 1999, in his sleep at the age of 70, suffering a heart attack.

Friedrich Christian Anton "Fritz" Lang

The name of Friedrich Christian Anton "Fritz" Lang will always be remembered when talking about the history of science fiction films as he was one of the few directors who made science fiction films in the initial years of the black and white era.

Friedrich Christian Anton "Fritz" Lang (December 5, 1890 – August 2, 1976) was an Austrian-German-American filmmaker, screenwriter.,

Lang came from a Moravian descent. His Jewish parents had converted to Roman Catholicism before Fritz's birth. Lang frequently had Catholic-influenced themes in his films. Although an atheist, Lang believed that religion was important for teaching ethics.

After finishing school, Lang briefly attended the Technical University of Vienna, where he studied civil engineering and eventually switched to art. In 1913, he studied painting in Paris.

At the outbreak of World War I, Lang returned to Vienna and volunteered for military service in the Austrian army and fought in Russia and Romania where he lost sight in his right eye.

He was discharged from the army with the rank of lieutenant in 1918 and later hired as a writer at Decla Film, Erich Pommer's Berlin-based production company.

He became director at the German film studio UFA, and later Nero-Film, just as the Expressionist movement was building. In this first phase of his career, Lang sought to synthesise art cinema with popular thrillers.

His famous classic science fiction film Metropolis was written 1927 which achieved cult status particularly for its technical finesse and use of massive stylized sets with 25,000 extras an epitome of German expressionism. It was followed by another science fiction film Woman in the Moon in 1929. The films flopped financially and bringing down the company UFA which was bought by a right wing businessman and politician.

In 1931, independent producer Seymour Nebenzahl hired Lang to direct M for Nero-Film. His first "talking" picture, considered by many film scholars to be a masterpiece of the early sound era, M is a disturbing story of a child murderer who is hunted down and brought to justice by Berlin's criminal underworld.

He made films on psychological conflict, paranoia, fate and moral ambiguity.

His film The Testament of Dr. Mabusewas cancelled by Joseph Goebbels, on the plea that it posed a threat to public health and safety. Goebbels stated that the film "showed that an extremely dedicated group of people are perfectly capable of overthrowing any state with violence",

Lang sold his wife's jewelry and fled by train to Paris. Later he went to America and worked for major studios and made more than 20 films. One of Lang's most praised film is the police drama The Big Heat (1953), known for its uncompromising brutality, especially for a scene in which Lee Marvin throws scalding coffee on Gloria Grahame's face.

Interested in remaking The Indian Tomb , Lang returned to Germany to make his The Tiger of Eschnapur and The Indian Tomb.

On February 8, 1960, Lang received a star on the Hollywood Walk of Fame for his contributions to the motion picture industry. Lang died from a stroke in 1976.

CHAPTER 6
BEST HOLLYWOOD SCIENCE FICTION FILMS

Avatar

It is a matter of pride for science fiction movie fans that the second highest grossing movie worldwide in the history of cinema is a science fiction film Avatar. . It has been written, produced and directed by James Cameron with a $ 300 million budget starring Sam Worthington, Zoe Saldana, Stephen Lang, Michelle Rodriguez, and Sigourney Weaver. It won two Golden Globe awards.

Avatar is clearly one of the few science fiction movies which places environmental concerns central to its theme. Its success primarily lies in the fact that it is a complete CGI based movie with virtually every detail rendered on computer.

The detailed intricacies of the flora and fauna of an imagined world may not have been possible to create in any other way if not digitally. The lush bright fluroscent colours of the environment, the mountains, the exotic plants and animals makes the audience completely lose himself/herself in the fantastic world.

The long blue and spiry Na'vi with their elongated ears initially may look odd but we slowly get used to them. The story has all the ingredients of typical Hollywood adventure and romance with strong environmental theme and a throw back on the rapacious colonial exploitation of natives.

The film talks of a time when humanity is faced with a global energy crisis having exhausted Earth's natural resources, and wants to mine energy rich element unobtanium, found on the moon Pandora, a jungle planet orbiting a gas giant in the distant Alpha Centauri galaxy. Pandora's atmosphere is toxic to humans and the Na'vi, a blue coloured race of beings, already inhabit it and protect its environment vigilantly.

Scientists have managed to create avatars, half-human, half-Na'vi hybrid clones that can freely move around and operate in Pandora's poisonous atmosphere. Human operators have their consciousness "uploaded" into these surrogate bodies.

Jack Sully a paraplegic marine is given a second chance at life when he replaces his identical twin brother as an avatar operator. He is shipped to Pandora to serve as a bodyguard for Dr. Grace Augustine, head of the Avatar program, and xenobiologist Dr. Norm Spellman as they study the native flora and fauna of the planet.

Jake is to learn more about how to extract the unobtanium and encourage the Na'vi to leave their "Hometree," or through force, if necessary. A large alien animal attacks Jake while he is on duty protecting the two scientists, and he runs into the jungle to escape it. There, a female Na'vi named Neytiri rescues him and takes him to her tribe after witnessing an omen that seems to suggest that Jake is an asset to the tribe. Neytiri's mother and spiritual leader, Mo'at, orders her to induct Jake into their clan. When Jake returns to headquarters and human life, Colonel Quaritch, Security Head for Resources Development Administration (RDA), promises to give Jake back his ability to walk if he agrees to spy on the Na'vi and collect information about their customs and secrets.

Jake soon becomes supportive of the Na'vi's plight as he gains a better understanding of their ways, and he proves himself to be a worthy Na'vi warrior. He is eventually integrated into the tribe and Neytiri chooses him as a mate. Jake now supporting the Na'vi sabotages a bulldozer that was tasked with destroying a sacred site, an act that prompts the RDA to order the complete destruction of Hometree.

Dr. Augustine argues that the destruction of Hometree would also affect all life on the planet, so Parker Selfridge, RDA Head Administrator, gives her and Jake one hour to plead their case with the Na'vi to evacuate their homes.

Back with the Na'vi, Jake acknowledges that he initially worked as a spy, which makes him a disgrace in the community. Colonel Quaritch has Jake, Dr. Augustine, and Norm unlinked from their avatars and thrown in a prison cell, but they are able to escape with the help of a sympathetic pilot named Trudy. During their escape, Dr. Augustine is fatally injured.

Wishing to win back favor with the Na'vi, Jake sets into motion a daring plan to connect his mind to a toruk, an immense dragon-like creature both revered and feared by the Navi, in order to redeem himself with the tribe. When he successfully binds with the toruk and becomes its rider, he wins the respect of the Na'vi once more. He manages to locate the survivors of the Hometree destruction at the Tree of Souls and begs Mo'at to heal Dr. Augustine. They attempt to "upload" her consciousness into her avatar body with the aid of the Tree of Souls, but her human body dies before the process is completed.

Jake, gathers and unites the remaining tribes to fight the RDA . Quaritch mounts a pre-emptive strike anticipating the retaliation. He targets the Tree of Souls, thinking that its destruction will dishearten the Na'vi.

The RDA's superior technology and weaponry overwhelm the Na'vi, but the animals of Pandora join in the fight, and change the course of the battle.

Jake and Quaritch face off and Jake manages to destroy the bomber craft before it can incinerate the Tree of Souls. Quaritch escapes and severs the avatar link unit that houses Jake's human body, subjecting him to Pandoran atmospheric toxins. Neytiri arrives just in time, however, killing Quaritch and saving Jake.

All humans, save for Jake and a select few, are commanded to leave Pandora and return to Earth. Jake's consciousness is then permanently transferred into his avatar.

The idyllic and visually stunning Pandoran world is created with attention to the smallest detail. It seems so real, so beautiful and exotic that one is automatically drawn into its spell. Its simple innocent Nature worshipping inhabitants draw our love and sympathy rather than the arrogant military commanding officers of the Resources Development Administration out to destroy the environment of the world of Pandora in their rapacious need for mining the mineral Unobtanium.

Even though talking about a distant time in the future and a world light years away from earth , the issue is relevant and topical today because even today the existence of large number of indigenous people, tribals and their local culture customs and traditions and languages all over the world face the threat of being wiped out by commercial corporate business interests.

The whole moral question being asked in the film is whether humans have the right to use military might and technology to extract valuable natural resources for their own need or profit by destroying or displacing the indigenous original owners of that valuable resource.

Although the movie is a science fiction story set in the future (year 2154) and the "native" Na'vi people on the planet Pandora have blue skin, the resemblance to the colonial exploitation by the white people of the black/brown/yellow indigenous natives all over the world is writ large. There is obviously a strong anti-war, anti-colonialism, and pro-environmentalist message in the whole movie making it more relevant.

The anti war messaging is very clear with RDA war machine painted as the evil destroyer, with words like "preemptive strikes," "fighting terror with terror" reminding one of Bush Administration's war against international terrorism after 9/11 and helicopter gunships flying low above the jungles reminding one of Vietnam war.

The white male character becoming the leader of the "colored" people in their own struggle, has come for criticism as also why story is narrated from the white character's point of view.

Even though critics have derided the typical motif of white male in love with native girl finally standing against his own kind as being very commonplace, the movie stands out not that much for its theme but more for creating an amazing three dimensional strikingly beautiful and colourful imaginary world which is complete in itself in minutest detail with people, floraand fauna ,it own geography and history ,myths and magic. It is unlike anything which audiences had seen before.

Blade Runner

Ridley Scott's Blade Runner (1982) as a science fiction film is visually so powerful that its setting is supposed to have been an inspiration for a number of movie makers, directors , cinematographers, who were looking for the right kind of feel and atmosphere for futuristic settings and locales.

But Blade Runner's cult status may not be only be due to its visuals but its engagement with some of the deep philosophical questions which are becoming more and more relevant to us as we progress technologically.

Actually Scott's visualization of the cyber punk world into the cinematic idiom was really liked by directors, many of whom followed his visual themes.

The twenty first century led to a new kind of science fiction becoming popular known as cyber punk which basically was a combination of hi tech and low life. It depicted future society as we advance technologically with artificial intelligence, robotics, super computers, slowly gaining positions of power in socio-economic and political structures and old civilizational values unable to cope up with the technological advances. The relentless march of technology and progress leads to not only the growth of new authority figures controlling the huge techno-corporates but also leads to the sprouting of counter culture groups of rebels, the anti authority, anti socials - the cyber punks in a dysfunctional society.

The whole concept of cyber punk in science fiction shows a futuristic technologically advanced dystopian society. There is high level of technology but it has bred and led to social chaos, rebellion and degeneration rather than social cohesion and regeneration.

"Blade Runner" was able to create the perfect visual depiction of such a society, the look, the atmosphere and world of cyber punk which the sci-fi directors and cinematographers were hunting for.

The highly technological, inhuman, dictatorial and oppressive world rampant with low life and anti social elements.

The 1982 film Blade Runner created a flashy, gaudy, plastic world full of neon signs and lights in urban settings reflecting the crass commercial artificial life. The use of neon colours and lights represented urban nightlife and streets.

Use of huge Xenon spot lights (used in sports events)to pinpoint and brightly lit up the interiors demonstrated how hovering airships monitoring all movements could sharply intrude into the privacy of the citizens in the futuristic scenario using the strong spotlights from above.

Besides use of lights, use of smoke and rain in the city atmosphere further enhanced the play of light and darkness in the myriad of watery reflections. This also visually highlighted the major philosophical theme of the film - truth and falsehood. The film deals with real humans and genetically engineered humans called replicants. The replicants look and behave just like humans. How do we know who is the true human and who is the replicant ?

Blade Runner is based on the science fiction novel by Phillip K. Dick 'Do Androids Dream of Electric Sheep?' Blade Runner is written by Hampton Fancher and David Peoples it stars Harrison Ford, Rutger Hauer, Sean Young, and Edward James Olmos.

The film is set in a dystopian future Los Angeles of 2019.

In a cyberpunk vision, humanoid androids with short, fixed lifespans have been manufactured - which are illegal on Earth, but are used in the off-world colonies ,as slaves for manual labor.

Four rebel replicants after mutiny on an off-Earth colony, revolt and after killing many humans and commandeering a spaceship land on earth to try to enhance their lifespan after getting in touch with their creator, Dr Eldon Tyrell, head of Tyrell Corporation.

The hero Deckard is a Blade Runner, a law enforcement officer, who is supposed to locate and kill replicants (genetically engineered human-like creatures) who have a life span of four years. Deckard who is called back from retirement tasked to kill the four replicants visits the Tyrell Corporation, manufacturers of the replicants, and meets Rachael, a worker there who is unaware that she is a replicant herself.

When Rachel comes to know that she is a Replicant she too absconds. Deckard discovers and kills three of the rebel replicants and also locates and shelters Rachael whom he starts loving. The only rebel Replicant left Roy Batty learns from Tyrell (founder of the corporation) that his lifespan cannot be extended. Tyrell consoles him by stating that "the light that burns twice as bright burns half as long, and you have burned so very very brightly"

Roy kills Tyrell.

Towards the end of the film Roy dies as his term ends but acting out of his character and showing empathy Roy saves Deckard.

The film ends with Rachael and Deckard making their escape together.

The questions what is the difference between humsn and robot become highly relevant in today's world as we continue to progress in creation of high performance robots taking on many of the tasks performed by humans. How we humans are going to deal with robots as we create them more and more like ourselves. As creators will we have the right to exploit them, enslave them ,kill them? .

Should similar set of ethics and laws apply to both ? What kind of legal rights should robots, cyborgs, androids have?

The questions become particularly critical when we start implanting robots with "memories" like Rachel in Blade Runner to create a sense of personhood. Again as the memories actually belongs to someone else (Tyrell's niece). Is Rachel Tyrell's niece ? Are we not the sum of our memories? But then childhood memories of Rachel can never be of Rachel because she was created in a laboratory as an adult and never had a childhood.

Deckard's identity also becomes ambiguous in the film and there is no clarity as far as whether he is a Replicant or human because he does not answer Rachel's question whether he had taken the Voight-Kampff test.

At a deeper level the film is rich with philosophical complexities but at the surface level also one can enjoy it as a tense and absorbing action drama. It is these multiple levels of interpretation which makes the film universal and a masterpiece.

Extra Terrestrial

Extra Terrestrial clearly is one of the most emotionally touching science fiction movie about the friendship of a little boy from earth with an alien from outer space. The deep love and affection which grows between two young living things seems to go beyond any difference they have as a species and that too belonging to two different parts of the universe. Love and compassion seems to cross all physical boundaries.

The ET gets left behind when humans come searching the area with flashlights and the spaceship is forced to take off in a rush.

The movie's innovation lies in the fact that instead of having the usual terrorizing fearful aliens coming to destroy , you have friendly , delicate extra terrestrial abandoned in ferocious world (at least as it seems to the little one with predators like dogs and raccoons and also naughty boys and automobiles).

Henry Thomas plays the little boy Elliot who befriends ET and Robert MacNaughton (older brother Michael), and Drew Barrymore (kid sister Gertie).

He is able to cajole ET out hiding and his natural fear of humans. The E.T. is seen to be almost like a curious human child prone to misadventure. Televisions, telephones, refrigerators all excite him with their newness and he wants try these out.

With powers of telepathy and telekinesis he communicates with Elliott. His desperation to go home is palpable when he again and again points his long thick stubby finger to the sky towards the planet where lies his home .

The film lets the audience know about extra-terrestrial medicine, biology and communication.

The ET is naturally the hero of the movie with its extremely cute walking style, look and laughter besides imitations of human speech. A creature created in such a fashion that the audience would soon fall in love with him. Its rare for a science fiction film to be made which is just restricted to Elliot's home and his bedroom and the suburban neighbourhood.

One of the most iconic moment of the film is Elliot and E.T.'s flight through the air on Elliot's bicycle which somehow seems magical with its beauty as their silhouette moves across an enchanting full moon and leaves us with a feeling of "wow"

E.T. is portrayed as an intergalactic botanist, a vegetarian whose very touch can heal the wounded and bring dying plants back to life. In contrast to ET are the government agents for whom ET is just a specimen. They do not seem to have the compassion,intuitive understanding or psychic power to connect with the ET.

Produced and directed by Steven Spielberg, and written Melissa Mathison the whole movie is shown through the view point of the children and even shots are taken from low angles, reflecting the point of view of the young protagonist and his child-sized alien friend.

Critics have also given religious connotation to ET as a kind of Messiah with healing touch and words of wisdom with the message of love and peace.

The 1982 film was a huge commercial success earning $793 million, or nearly 80 times its production cost of $10.5 million. It also achieved critical acclaim, winning four for Oscars for Best Visual Effects, Best Sound Effects Editing, Best Sound and Best Film Score.

It stands as one of the greatest films ever made with generations of filmgoers giving it the thumbs up. It has been selected for preservation by the National Film Registry in the US .

The young boy Henry Thomas got the role when he literally cried during audition where he was asked to play the role of Elliot trying to stop ET being taken away by government agents.

The fascinating look of ET was developed from a painting by special effects artist Carlo Rambaldi(who had designed the mechanical head effects for the xenomorph in Ridley Scott's Alien and the visitors from Spielberg's own Close Encounters of the Third Kind

The basics were all there with a long neck, an oblong head, and large eyes. To make the alien a lovable creature Spielberg had Rambaldi study photos of elderly people who lived during the Great Depression and famous people like Albert Einstein, Ernest Hemingway, and Carl Sandburg, hints of their visage maybe seen in ET's facial characteristics.

The actual animatroni ET was a robotic puppet with puppeteers observing and controlling commands of the puppet's movements remotely looking at TV monitors. ET's hand movements were actually that of a mime artist.

The music was composed and conducted by John Williams. The music is deep and orchestral. Instruments like trumpet, violins, cellos and other strings instruments besides drum beats and brass instruments are used to build up the tempo in relation to the dramatic action taking place in the film building up the tension and the excitement.

Nominated in nine categories at the 1983 Academy Awards, including Best Picture and Best Director, the film won four Oscars, for Best Sound Effects Editing, Best Visual Effects, Best Original Score and Best Sound.

John Williams sound and music throughout played a crucial role in enhancing the impact of the movie. Music was completely fused into theme, character and storyline of ET as an integral part without which ET would not be what it is.

He captures the mood and emotional drama of every scene by the accompanying sound and music to create the right atmosphere.

The Flying Theme played against the night sky's full moon as ET and Elliot bicycle across, and at sunset as number of young children on bicycles one after the other cross the round shining yellow globe of the sun is iconic. The film's ending with departure scene is accompanied by unforgettable music with use of horns, resounding full orchestral sounds with flourish and rolls adding to the grandeur of the ET going away to his own planet in the alien spaceship.

Fantastic Voyage

In 1966 came out the landmark film 'Fantastic Voyage' for which the advertisements at that time had declared tantalizingly "You've never been HERE before!"

It was a riveting science fiction story about a group of surgeons/scientists being miniaturized along with their submarine and injected into the blood stream of an injured scientist to dislodge a clot in his brain.

The story has scientist Dr. Jan Benes inside Soviet Union during Cold War days plans to defect and escapes with help of American intelligence agents but injured during an attempt to assassinate him and develops a clot in his brain which has to be removed. This can only be done from inside the body and a rescue team is formed by CMDF(Combined Miniaturization Deterrent Forces) which is given the task.

American and Russian scientists both have been working on a top secret method to miniaturize both humans and objects . While American scientists can only do it for one hour , Dr Jan Benes knows how to do it indefinitely. So there is strategic importance to revive Dr Jan Benes who holds the secret of the procedure. The miniaturization technology holds enormous possibilities for the military as it could help transport Defence forces and armoured vehicles stealthily in miniaturized form behind enemy lines to be de-miniaturized there.

Fantastic Voyage is a taut and tense drama as the rescue team of surgeons and crew has just one hour to perform their task because they would stay miniaturized only for an hour.

The excitement of the film starring Stephen Boyd, Raquel Welch, and Donald Pleasence is further enhanced by the presence among the crew of a saboteur (owing allegiance to the other side), who is there to see that the mission fails.

Dr Jan Benes has been placed in deep hypothermia - as low as compatible for sustaining human life: his heart rate has been reduced to only 32 beats a minute; his respirations have been reduced to only 6 per minute.

The submarine and the crew within are miniaturized into an ampule that is hypodermically injected into the patient's carotid artery.

There are tense moments when the submarine the passes through the inner ear via the endolymphatic duct, where any sound emanating from the operating room could create large scale vibrations to topple the submarine .

The submarine and its crew are foreign body inside the body and therefore the natural tendency will be for the bodies defense mechanism to destroy it. Therefore going out of the submarine is also risky and the crew members also face attack from anti bodies.

The submarine encounters an abnormal arteriovenous fistula which is abnormal connection between an artery and a vein in which the blood flows directly from an artery into a vein, bypassing some capillaries. This leads the submarine into a whirlpool and directly into the jugular vein and there onto the superior vena cava.

It is really fascinating and gripping to see the interiors of the body shown from the perspective of the travelling scientists as they go through the arteries and veins, through the heart of the patient whose beating has to be stopped for sometime as long as they are in it, so that there submarine is not crushed by the pressure of the heart's pumping.

The submarine travels through the dark domed cave like heart through the vena cava and via the tricuspid valve to the right ventricle and somehow it gets into the pulmonary artery and reaches the lungs.

The ingenuity of the scientists is shown by their successful attempt of collecting oxygen directly from the alveolis in the lung walls, by a snorkel.

Through the lymphatic system and the subarachnoid cavity at the base of the brain they go carefully into the brain avoiding important nerves so that no damage will occur to the brain.

The surgeons wearing scuba gear swim in the brain to reach the clot which is dissolved with the laser gun.

Dr Michaels turns out to be the saboteur and attempts to crash the submarine in the clot area to kill DR Benes. However a crew members fires laser gun at the submarine damaging it and it crashes. A white blood corpuscle finally gobbles up Michael and the wrecked submarine.

As time is running out the crew have to make a dash for it by swimming along the optic nerve, out through the tear duct, to the cornea and to the lower eyelid. A glass slide is held to catch the crew in the tear drop. It is a nail biting ending because any delay would have led to the crew gaining back their own real physical size leading to the death of the scientist. Interestingly members of operation communicate with the outside world using Morse code.

Though not getting involved in any deep seated philosophical discussion on the wonders of the body. The major view points are definitely put forward through the physicians Dr Peter Duval and Dr Michaels (saboteur), one is the spiritual and the other is the scientific.

Duval is clearly amazed by the automatic process of oxygenation in the lungs where the corpuscles undergo the exchange of gases and takes it to be a "miracle" demonstrating the presence of "creative intelligence".

Dr Michaels, who is the villain, rejects the view and points out that it is process is just part of the development due to the evolutionary process clearly the scientific view point. Michael later taunts Duval and wants him to look for the 'soul' when the submarine reaches the brain.

What is greatly appreciated in the film is the display of detailed living and functioning interior anatomy of the body and its contents which is truly amazing and full of wonder and awe. Therefore whatever the villain Dr Michael might say the movie itself does leave the audience wondering, " Who is the Creator of this marvelous piece of intricately fashioned and perfectly functioning human body?"

The film was produced at a time when visual access to the body's interior was extremely limited even to medical professionals. The Screenplay was done by Harry Kleiner and was based on story by Otto Klement and Jay Lewis Bixby. Novelization of the film written by the science fiction writer Isaac Asimov was published in 1966

For the Fantastic Voyage the most expensive sets which took more than a year to build were created costing a whopping 6.5 million US dollars at that point of time.

A submarine travelling with surgeons and medical team in the carotid artery of the patient on the operation table is unquestionably thrilling medical sci fi.

Today as nano technology is making great strides medical procedures being conducted by microrobots in the tiny realms inside our body is slowly becoming a reality.

Battery-powered gadgets like capsule-sized camera that can be swallowed in order to capture video images of the esophagus, intestine, and colon are being introduced.

But in 1966 when many of the high tech medical gadgets had not been manufactured to have a real time three dimensional view of the body's interior was not possible , one can imagine the visual impact, the thrill and the excitement of looking at a submarine carrying surgeons moving in the bloodstream and miniscule surgeons wearing scuba gear swimming inside the brain to reach the clot to perform surgery.

Director Richard O. Fleischer who made this film has made many memorable films including 'The Narrow Margin', '20,000 Leagues Under the Sea' , 'Doctor Dolittle', 'Tora! Tora! Tora.'

The film 'Fantastic Voyage' was the first movie taking a journey inside the body and was described at the time of its release as "an enlightening excursion through inner space - the body of a man." and being "... imaginative, inventive, ingenious...."

There was obviously no CGI and the entire submarine had to be built both interior and exterior, actors were suspended by wires and a lot of effort was taken hide those contraptions. Many different size of models were prepared as shots kept shifting from life size in the real world to that of miniscule size inside the body

Matte paintings were used to a great deal to graphically display the submarine and its crew moving through the blood stream with red corpuscles,whitecorpuscles,antibodies, plasma and platelets and other bio-parts.

Different kinds of gelatin,vaseline and fibers and oils were used to create the look of body fluids .

Academy of Motion Picture Arts and Sciences voters acknowledged the picture's marvels with the Oscar for Best Art Direction - Set Decoration, Color; as well as the Oscar for Best Effects, Special Visual Effects.

For many years clips from the film was shown in medical schools to illustrate human anatomy and physiology.

Gravity

One of the most chilling and realistic portrayal of disaster striking in space is seen in the film Gravity.

The greatness of Alfonso Cuarón's "Gravity" lies in the fact that he is able to show the stark contrast of the fragile and puny astronauts in front of the vast desolate expanse of unending space all around.

The vulnerability and helplessness of astronauts and their craft is established early on in the film when it is damaged by a meteor shower. In front of the immense destructive power of Nature in the wide expanse of space all attempts by Man seems to be feeble and futile.

No wonder there is a strong spiritual undercurrent in the film which highlights the fact despite Nature hurling its utmost fury how miraculously Man is saved and eventually brought to safety.

As a child, Cuaron had been greatly influenced by the 1969 about three astronauts who become stranded in space and subsequent rescue by the Soviets which he saw many times.

In the space disaster film as a female lead Sandra Bullock (Dr Ryan Stone) the heroine, has perhaps given one of her best performances in the film as an astronaut displaying her range of histrionic skills from a scared vulnerable and dejected woman to one who undergoes every challenge with courage grit and determination.

She virtually carries the whole movie on her own shoulders which won several Oscars including Best Director for Cuaron, Best Cinematography, Best Film Editing, Best Music (Original Score) , Best Sound Editing, Best Sound Mixing and Best Visual Effects.

Dr. Ryan Stone is on a spacewalk repairing a panel on the Hubble Space Telescope with commanding officer astronaut Matt Kowalski (George Clooney) when Mission Control instructs them to abort their spacewalk and return to their space shuttle Explorer as a large debris is coming towards them and they lose contact.

High-speed debris breaks the robot arm off from the STS and Stone is forced to release the tether holding her to the robot arm. Once free, she is thrown far from the STS. As she is panic stricken Kowalski wearing a thruster pack comes and attaches her tether to him.

As they go towards STS Explorer they find the dead body of an engineer and when they reach STS Explorer they are shocked to find the craft ripped open and all the astronauts and engineers dead.

Kowalski decides to reach International Space Station (ISS) 100 Km away using thruster packs before debris returns in 90 minutes in its completing its orbit . With Stone's oxygen reserve running low, they float through space to the ISS. Kowalski asks Stone about her life back home and learns that her daughter died in a schoolyard accident.

The ISS crew has already left and the craft is useless for return to earth as its parachute has been already deployed accidentally.

Kowalski plans to use it to travel to Tiangong, the nearby Chinese space station, to retrieve another module that can take them to Earth.

While maneuvering fuel for the thruster pack the tether holding them together breaks and Stone's leg becomes entangled in the Soyuz's parachute lines. Stone grabs Kowalski's tether, just barely stopping him from flying off into space. But Kowalski realizes that both of them are likely to be carried away and decouples his end of the tether so that Stone can survive.

Kowalski floats away to a lonely end but radios her with additional instructions about how to get to the Chinese space station, encouraging her to continue her survival mission.Stone enters the ISS through an airlock and gets out of the spacesuit but has to face a fire and despite her attempts to douse the flames eventually has to dump ISS and separates from it in the Souyze, while ISS is completely destroyed.

Her woes are not yet over as she find out when she aligns her craft towards Tiangong the Chinese spacecraft that the tanks are empty. She is numbed and in panic believing that her end has come.

She slows the oxygen flow which will cause her to fall into unconsciousness from lack of oxygen before she dies. As she begins to lose consciousness, Kowalski suddenly appears outside, opens the spacecraft lock, and enters the cabin. He cheerfully asks her if she wants to live or die and tells her to use the Soyuz's landing rockets to reach Tiangong.

What Kowalski tells her is truly motivating, "I get it, it's nice up here. You can just shut down all the systems, turn out all the lights… and just close your eyes and tune out everybody. There's nobody up here that can hurt you. It's safe. I mean, what's the point of going on? What's the point of living? Your kid died. Doesn't get any rougher than that. But still, it's a matter of what you do now. If you decide to go, then you gotta just get on with it. Sit back, enjoy the ride. You gottaplant both your feet on the ground and start livin' life… Hey, Ryan? It's time to go home."

She gains full consciousness and finds that Kowalski's appearance was a hallucination. Was it some Superior being, was it Kowalski's spirit ? The movie does not make it clear but the message has a transformative , regenerative , impact on Stone. She wants to move on. The past is a closed chapter. She has been psychologically reborn.

She restores oxygen and uses the landing thrusters to navigate towards Tiangong. As she does not have any way to slow her movement to dock with the Chinese station , she ejects herself from the Soyuz taking recourse to the risky method of explosive decompression and pressure in the fire extinguisher to push herself towards the abandoned Chinese craft Tiangong.

Stone is unable to separate the capsule from the space station as Tiangong has also been damaged from the debris hits. The space station breaks up and the capsule separates , Stone fires the device that separates the capsule from the rest of the spacecraft and aims towards earth. As it descends through the atmosphere , Mission Control informs on the radio that rescue is on its way.

The chute automatically deploys and Stone lands in a lake near the shore. The cabin is full of smoke, and after she blows the hatch the capsule tilts, allowing water to enter. Just when she thought she was safe, she's unable to exit the capsule. The capsule sinks to the bottom with Stone inside. Stone finds a bubble of air inside the capsule and exits the capsule, shedding her spacesuit so she can get to the surface. She swims to shore and with difficulty, gets up and takes a few shaky steps.

The realism of the film and the acting virtually makes the audience live through her experience. One of the few films which does not seem to be science fiction but real. There is actually no sound when the perspective is from space. All the spaceships are modelled with accuracy and great precision and attention to detail again making you seem you are in space.

The shocking fearful ordeals one after the other which Stone undergoes underlines the cruel and unforgiving nature of the vast wilderness of space. Physically, mentally, emotionally, psychologically, spiritualy Stone is completely drained out in her determined struggle for survival and survive she does.

The film's message is very loud and clear that Man and his spacecraft are truly puny , delicate , weak in front of Nature in its fury, in its cold and fierce aspect in the vast desolation of Space. Man would definitely lose out but Man is not alone, as events pan out for Stone there seems to be some some superior power which we may not be able to see but which issheltering and protecting her ready to help and guide and save her from all kinds of troubles. There is a silver lining in every dark cloud.

Innumerable times when she has lost all hopes Stone finds some solution, some ingenuous technique, and some way out of her predicament. Its as if some Saviour is at hand who is watching her with love and care at every point of crisis. There is a time when Stone is completely dejected , depressed and at loss to the point of attempting to ending her own life.

She reduces oxygen and is about to give up her life having lost all hope when that the superior power seems to take over. She has a hallucination of Kowalski revealing a way out, following which she is again optimistic and determined to make an attempt to return to earth. Somehow this encounter also brings about a closure to her continued dejection and depression following the death of her daughter.

In the dreamy hallucination where she has the transcendent vision she asked Kowalski to communicate her love to her four year old daughter. The cocoon in which she is supposed to have enclosed herself at least psychologically seems to have been broken. Its in a way her re-birth as she was almost dying (going to commit suicide)before she has the hallucination.

This re-birth metaphor is further re emphasized forcefully in the last scene of the movie.

The fact that the film has a spiritual religio undertone is also seen in the kind of symbols being played up like metaphor of rebirth where Ryan gets into an airlock and curls into the fetal position while the parachute chord floats near her like umbilical cord, reference to the glint of Sun in the Ganges, the St. Christopher icon on the instrument panel of the pod,the statue of Buddha in Shenzou capsule of the Chinese Tiangong Station.

The film points out that despite all the technological developments there is another spiritual dimension of reality which cannot be completely negated almost harking back to the 2001 Space Odyssey made decades earlier.

Cuaron has said that "*2001: A Space Odyssey*, it's clear hat its ghost was haunting me.

"I was aware of the image of Ryan Stone in fetal position because rebirth was a main theme of the film, but isn't rebirth also the theme of *2001: A Space Odyssey* with its beautiful ending of the star child?"

"The sparse dialogue and use of silence in *Gravity*, it's not only a legacy of *2001*, it is a legacy of all Kubrick's work that, like many other film masters, believe in the power of cinema as an experience that can convey themes through its own language"

It is the spiritual dimension which aids us when confronted with desolation in wilderness of Space or with psychological desolation following death of someone close like Stone's daughter.

While the entire film is shown to be taking place in Space with only dark black and white dominating, its stunning impact is due to the sudden appearance of vivid colours of brown and green and blue of the earth at the end of the film. Secondly the last scene is shot entirely on a 65 mm film with greater sharpness and details unlike the entire film which is shot digitally.

Director Curaon has best described the scene, "She's in these murky waters almost like an amniotic fluid or a primordial soup. In which you see amphibians swimming. She crawls out of the water, not unlike early creatures in evolution. And then she goes on all fours. And after going on all fours she's a bit curved until she is completely erect. It was the evolution of life in one, quick shot".

Stone crawls to shore, grasps the wet sand in her hands, and mutters the final word of the movie: "Thanks."

Having virtually lived through the harrowing experience of the movie the sense of relief and happiness having reached earth safely and the shining welcoming sight of the colourful earth itself makes it one the most refreshing and transformative ends to a movie.

The greatness of Gravity will always lie like every other great movie in the fact that while watching it we are not just made to go through moving images but go through a deeply moving experience of affirmation and trust.

Iron Man

The first Iron Man movie released in 2008, is science fiction movie and does not fall under the super hero or fantasy movie category though many may feel inclined to put this flying hero into that category..

Directed by Jon Favreau written by Mark Fergus & Hawk Ostby and Art Marcum & Matt Holloway it is high drama from the start to finish with a simple story line of how the hero Tony Stark (Robert Downey Jr) a rich charismatic, smart and techno savvy owner of a company with Pepper Potts (Gwyneth Paltrow) Stark's loyal aide, and Obadiah Stane (Jeff Bridges), Stark's business partner, manufacturing high-tech weaponry, transforms into Iron Man.

In a almost real life scenario Tony is kidnapped in Afghanistan where he had gone to demonstrate the "Jericho" missile by terrorist group Ten Rings who hold him in a cave.

In the cave he meets Professor Ho Yinsen who implants an electromagnet into Stark's chest to keep the shrapnel in his chest from reaching his heart and killing him. Both of them also build quietly an arc reactor to power the electromagnet and an armored suit.

However terrorists find out what they are up to and using the armored suit Stark escapes but the professor is killed by the terrorists. Stark destroys the weapons of the terrorists and flies out but crashes in the desert destroying the suit.

Back in his home he makes another much sleeker and powerful suit and arc reactor. He also decides that his company will not make weapons any longer displeasing company manager and partner of his father Obadiah Stane (Jeff Bridges).

Stark gets to know from a reporter that his company's weapons are still being made available to Ten Rings and they are using them to attack Gulmira the village of the professor Ho Yinsen.As Iron Man Stark wearing the suit flies to Afghanistan and defends villagers of Gulmira. On his return he is confronted by two airforce fighter jets and has to reveal his secret to Rhodes.

When Stark Industries holds a charity event, reporter Christine Everhart tells Stark that his company's weapons were recently delivered to the Ten Rings, and they are using them to attack Yinsen's home village of Gulmira.

Angry, Stark dons his Iron Man suit, and flies to Afghanistan to save the villagers. As he returns home, two fighter jets attack him. This forces Stark to reveal his new secret identity to Rhodes in order to end the attack.

By now, the Ten Rings have pieces of Stark's prototype suit, and they meet with Stane, who has been trafficking arms to criminals worldwide. He wants to replace Stark as Stark Industries' CEO, so he reverse engineers the wreckage of the prototype suit the Ten Rings gives him.

Stark has Pepper Potts hack into his company's database to track any illegal shipments his company made. She discovers that Stane hired the Ten Rings to kill Stark, but then they changed their mind. Potts then informs S.H.I.E.L.D. Agent Phil Coulson about Stane's anti national activities.

Stane can't duplicate Stark's arc reactor and wants to steal his. Wearing his Iron Monger armor, manufactured by Stane International and code-named I-M Mark One, with repulsor rays and an intense laser he attacks Stark who is under powered without his new reactor running at fully.

Stark gets Potts to overload the large arc reactor that powers his building which creates a huge power surge that Stane falls into the exploding reactor which instantly kills him.

Stark publicly admits to being Iron Man at the end of the movie a complete departure for such action heroes.

The film is directed by Jon Favreau; written by Mark Fergus, Hawk Ostby, Art Marcum and Matt Holloway. The film grossed $585 million worldwide and went on to make Robert Downey Jr. one of the best paid actors in Hollywood.

"When you talk about Marvel's rise, *Iron Man* set the tone of all of it," according to director of Avengers Joe Russo.

Favreau did not want to create a super hero character with whom the audiences could not identify with. He felt that sci-fi and superhero genre was full of heroes who look cool in action but did not emotively connect with the audience and "ultimately left audiences feeling cold". Tony Stark therefore needed to be more flesh and blood man facing all normal troubles and tribulations.

The courage and sacrifice shown by Iron Man's character contributed as much to the film's success as the special effects, according to its director.

The reason why Iron Man is science fiction and not a fictional or fantasy character is because all his power is supposed to be based on science and technology and not on magic. Even if some of the scientific products may not be available now they may be possible and some within foreseeable future. They are not products of magic or flights of fancy.

As Iron Man's name suggests, he wears a suit of "iron" which gives him his abilities—superhuman strength, flight and an arsenal of weapons—and protects him from harm.

In scientific parlance, the Iron man suit is an exoskeleton which is worn outside the body to enhance it.

In the 1960s, the first real powered exoskeleton appeared—a machine integrated with the human frame and movements which provided the wearer with 25 times his natural lifting capacity.

Controlling the exoskeleton with simple thoughts, may be available in future. Today already exoskeletons have been developed for medical applications, with devices designed to give mobility to amputees and paraplegics.

Work is also on to manufacture suits which make the weight almost 80 per cent lighter.

Eric Scott whom this writer has interviewed called 'Rocket Man' has flown 600 times on jet pack going upto 250 feet.

Director Favreauto create a seamless impact wanted to "strike a balance that mixes the visual effects with practical effects so audiences start to forget where one begins and the other ends.Sometimes that means cutting from one shot of Iron Man wearing the real suit to another that's completely a computer-generated image (CGI)."

All flying was not done with CGI and Motion Capture but there are many sequences of Robert Downey Jr. which were done with wire work particularly when he was flying around in his workshop in the studio by using a state of the art rig.

A film which is highly engrossing and visually stunning and a pure fun to watch sci fi got Oscar award for Visual Effects and Sound Editing.

Jurassic Park

Steven Spielberg's Jurassic Park must take the credit for making dinosaurs a reality for large number of movie goers all over the world particularly of the 1990s generation.

The creatures depicted in earlier films nor the stories had the kind of international impact this film had on the popular mind.

The point being made by Spielberg that don't mess with Nature was clearly driven home even though crowds packed the halls and went home after thoroughly enjoying their tense scary moments witnessing the huge T-rex on the 'screen'.

The story is a straightforward roller coaster adventure ride. Industrialist John Hammond has created a theme park of cloned dinosaurs by extracting dinosaur DNA from mosquitoes preserved in amber. DNA from frogs was used to fill in gaps in the genome of the dinosaurs. To prevent breeding, all the dinosaurs were made female.

Jurassic Park, is on an island and after a dinosaur handler is killed by a Velociraptor, the park's investors, represented by lawyer Donald Gennaro, demand that experts visit the park and certify its safety. Gennaro invites mathematician and chaos theorist Ian Malcolm, while Hammond invites paleontologist Dr. Alan Grant and paleobotanist Dr. Ellie Sattler to come and certify the safety of the park.

The group is shocked to see a live Brachiosaurus in the park while on tour of the park and are later joined by Hammond's grandchildren, Lex and Tim Murphy, while Hammond oversees the tour from the control room.

The tour does not go as planned, with most of the dinosaurs failing to appear and the group encountering a sick Triceratops.

As a tropical storm approaches mostof the park employees leave for the mainland on a boat while the visitors return to their electric tour vehicles, except Sattler, who stays behind with the park's veterinarian to study the Triceratops.

Jurassic Park's lead computer programmer, Dennis Nedry, has been bribed by Dodgson, a man working for Hammond's corporate rival, to steal fertilized dinosaur embryos. Nedry deactivates the park's security system to gain access to the embryo storage room and stores the embryos inside a container disguised as a shaving cream can.

Nedry's sabotage also cuts power to the tour vehicles, stranding them. Most of the park's electric fences are deactivated as well, allowing the park's Tyrannosaurus rex to escape and attack the group.

Grant, Lex and Tim escape, while the Tyrannosaurus injures Malcolm and eats up Gennaro. Nedry crashes his vehicle and is killed by a Dilophosaurus.

The survivors are cornered by the raptors, but they escape when the Tyrannosaurus appears and kills the raptors. Hammond arrives in a jeep with Malcolm, and the group boards a helicopter to leave the island.

Dinaosaurs have always fascinated Man and as far as science fiction films are concerned Jurassic Park stands out as one film which brought onto the silver screen the most believable and realistic dinosaurs for the first time.

The film was 1993 Oscar winner for Sound (Gary Summers, Gary Rydstrom,Shawn Murphy, Ron Judkins),Sound Effects Editing (Gary Rydstrom, Richard Hymns) and Visual Effects (Dennis Muren, Stan Winston ,Phil tippet, Michael Lantieri), and remained the highest grossing film of all time for many years . The hype and popularity of the film was such that nearly a thousand products were licensed with film logo from T-shirts to computers.

The earlier cinematic dinosaurs were no match for what was created by Jurassic Park which created a sense of awe and terror in the minds of the audience. Steven Speilberg's film is not just a monster movie but gives an important message through it which is that Man should not play God, a theme which has been recurrently occurring in many science fiction films earlier.

Biotechnology and genetic engineering are very powerful tools which must be handled with extreme caution and care.

We do not understand Nature fully and may not be able to do so even in the future therefore we must not tinker with Nature. Spielberg asks the fundamental question "DNA cloning may be viable, but is it acceptable?"

We cannot close our eyes to the dangers of science. The theme of Jurassic Park is clearly flagging the possible dangers of cloning particularly animals like the dinosaurs.

In a dialogue in the film by Malcolm is very pertinent where he says "Genetic power is the most awesome force the planet's ever seen, but, you wield it like a kid that's found his dad's gun."

The spectre and fear of large monsters created by genetic engineering gone wrong is hidden deep within human psyche which is brought out in the open in a film like Jurassic Park.

to create new organisms. Use of such power requires wisdom and patience. Malcolm punctuates his criticism in the same scene when he says, "Your scientists were so preoccupied with whether or not they could, they didn't stop to think if they should."

It is a nightmare through which the visitors to the Park go through. The relentless pursuit by the dinosaur who seem to be intelligent and kill so many people , the rest who flee for to save their lives are actually in no position to challenge them.

The technology, the machinery, the high voltage perimeter electric wire fencing actually is of no help because power had been cut off when the animals make their break through. The vehicles and guns are too puny including their buildings and shelters. Men have to literally run like rats to save their lives against these gigantic creatures.

Spielberg clearly wants us to have an attitude of reverence towards Nature which is again a very eastern pantheistic position. Malcolm says at one point "The lack of humility before Nature being displayed here, staggers me."

The main point of the movie seems that making interventions into Nature without thinking of possible repercussions can lead to very dangerous and uncontrollable results.No wonder genetic engineering is always shrouded in some controversy or another.

The science of genetic engineering which makes interventions into Nature must be based on very firm ethical base, seems to be the moral of Jurassic Park which also goes to make it a very relevant film for today's world.

The Matrix

The Matrix movie as a true work of art gives rise to several possible interpretations philosophical, sociological, economic. It was released in 1999 written and directed by the Wachowskis and produced by Joel Silver. It stars Keanu Reeves, Laurence Fishburne, Carrie-Anne Moss, Hugo Weaving, and Joe Pantoliano. In 1999, The Matrix got Oscar awards for Visual Effects (John Gaeta, Janek Sirrs, Steve Courtley, Jon Thum), Sound (John T. Reitz, Gregg Rudolff, David E. Campbell, David Lee), Sound Effects Editing (Dane A. Davis), Film Editing (Zach Staenberg).

The film shows Neo a computer hacker is contacted by underground freedom fighters who explain that Reality as he understands it is actually a complex computer simulation called the Matrix.

The Matrix has been created by a malevolent Artificial Intelligence to use people as an ongoing energy source while keep them engrossed in a simulated reality.

The leader of the freedom fighters Morpheus, believes Neo is "The One" who will lead humanity to freedom and overthrow the machines. Together with Trinity, Neo and Morpheus fight against the machine's enslavement of humanity as Neo begins to believe and accept his role as "The One".

Comic books like Japanese manga and anime, Hong Kong action and martial arts movies, and the counter-culture of '90s computer programmers, hackers and video game players have influenced the creation of Matrix.

William Gibson's novel *Neuromancer* relating the exploits of a data-thief influenced Watchowskis in creation of characters and how they interact with the digital world.

Wachowskis made Matrix a virtual world which all humans are plugged into and convinced into believing is the real world.

The Japanese animation film Ghost in the Shell, also had greatly influenced the directors.

Almost in a similar philosophical theme in Ghost in the Shell there is a vast network of AI connecting everyone and the protagonist a cyborg looking for the "Puppet Master" who can hack brains of cyborgs, wakes up and questions the reality around her, searching for a soul.

Director of Ghost in the Shell is supposed to have applauded the Wachowskis for being able to think like an animator and translate it into live-action.

Another film Dark City may have also influenced Matrix where the hero John Murdoch "wakes up" to the knowledge that his world isn't real and find agents unseen by most people.

The Matrix movie's greatest impact was the open debate it has led to on the possibility that the Reality as we know it may be false.

A doubt on which philosophers and thinkers have been cogitating since time immemorial including Plato and Descarte without coming to any definitive conclusion.

The movie as a science fiction suggests one possible scenario shockingly revealing that human beings are kept suspended in vats being exploited for bio-energy and kept occupied and happily engrossed in a simulated world. The Matrix is a computer-generated dream world designed to keep these humans under control.

Actually Plato had explored the idea that the real world is an illusion in the allegory of the cave in The Republic.

Plato imagines a cave in which people have been kept prisoner since birth. These people are bound in such a way that they can look only straight ahead, not behind them or to the side. On the wall in front of them, they can see flickering shadows in the shape of people, trees, and animals. Because these images are all they've ever seen, they believe these images constitute the real world. One day, a prisoner escapes his bonds. He looks behind him and sees that what he thought was the real world is actually an elaborate set of shadows, which free people create with statues and the light from a fire. The statues, he decides, are actually the real world, not the shadows.

Then he is freed from the cave altogether, and sees the actual world for the first time. He has a difficult time adjusting his eyes to the bright light of the sun, but eventually he does. Fully aware of true reality, he must return to the cave and try to teach others what he knows. The experience of this prisoner is a metaphor for the process by which rare human beings free themselves from the world of appearances and, with the help of philosophy, perceive the world truly.

Neo the hero of Matrix is pulled from a kind of cave in the first Matrix film, when he sees the real world for the first time. Everything he thought was real is only an illusion—much like the shadows on the cave walls and the statues that made the shadows were only copies of things in the real world. Plato insists that those who free themselves and come to perceive reality have a duty to return and teach others, and this holds true in the Matrix films as well, as Neo takes it upon himself to save humanity from widespread ignorance and acceptance of a false reality. Matrix also seems also to draw on French philosopher Rene Descartes writings who also strives to find out what is Reality ?

In Meditations Descartes after doubting the validity of all his senses for providing him knowledge of the truth about the real world affirms that the only thing about which he can affirm truthfully or without any kind of doubt , is the 'Doubter' or himself.

As long as he exists as a conscious being, he continues to be the truth and does not become nothing or false even though the entire world outside may be a false.

Thus the Matrix deals with very fundamental philosophical questions, the hero Neo learns about the Matrix, has doubts and fear, and eventually learns how to make a choice and alter his destiny.

Morpheus presents two pills, red and blue. If Neo selects the blue pill, he'll wake up again at home and remember nothing. If he selects the red pill, Morpheus will allow him to see the truth. One of the most powerful and visually shocking sequences in science fiction genre is clearly when Neo chooses the red pill and suddenly a mirror near him liquefies. When he touches it, its mercury-like substance oozes over him, threatening to envelop him. His world dissolves in front of him, and he panics. Neo wakes up naked and hairless in a vat of jelly, with plugs connecting him to the vat. Millions of similar vats, each with a human inside, stretch around him in all directions. Flying robotic insects drill a hole in the back of his neck. Then the jelly drains from Neo's vat, and Neo slips through pipes down into a pit full of water. A metal claw rattles down from a spaceship and plucks him up into the light.

Neo sees the Reality as it is.

Another philosopher from which The Matrix seems to have been inspired is Jean Baudrillard's Simulacra and Simulation which book is also seen in one of the shots in the film.Neo keeps his software inside a hollowed-out copy of the book Simulacra and Simulation .

Originally published in 1981, Baudrillard's book argues that late-twentieth-century consumer culture is a world in which simulations or imitations of reality have become more real than reality itself, a condition he describes as the "hyper-real."

Baudrillard argues that consumer culture has evolved in such a way that our lives are filled with simulations. In such a situation, the world of simulations increasingly takes on a life of its own, and reality itself erodes to the point that it becomes a desert. Morpheus introduces Neo to the real world by welcoming him to "the desert of the real," a phrase taken from the first page of Simulacra and Simulation.

The entire concept of the Matrix films can be interpreted as a criticism of the unreal consumer culture we live in, a culture that may be distracting us from the reality that we are being exploited by someone or something, some company or industry, just as the machines exploit the humans in the Matrix for bioelectricity. Our desires are inflamed by continuous messaging to buy products for consumption both for our body and mind. We do not get an opportunity to think rationally for ourselves whether we actually need the product. The need has already been created in us.

The producer or the product and service continue to exploit us as it is they who decide what it means to have a good time, a great product, a great relationship. An unreal world of the "smart and beautiful" is projected in front of the gullible consumer just so that he parts with his hard-earned money to continuously keep buying their product or service.

Karl Marx argued that the working class is exploited by the ruling classes, but the working class's exploitation is only possible because it does not perceive itself as being exploited. The working class misunderstands its own position because it is confused and distracted by social messages that give workers a distorted explanation of how they fit into the world—for example, religion, school, and ideologies such as nationalism and patriotism.

Neo's decision to fight and reveal the truth to all is the clarion call for action. The film does not believe that humanity enveloped under simulation cannot seek to get out of that situation by individual effort.

It is here we may also say that The Matrix at least for the Indian audiences may seem very familiar to the Indian philosophical concept of Maya in the Hindu philosophy. The vendantic philosophic idea that individuals are under the spell of Maya an illusion due to which they see this world and themselves as individual participants, whereas in reality they are one with the Supreme Being.

Enlightenment is achieved when one awakens from the illusory world (maya) and becomes one with the Bramhan, the Absolute, which is the truth.

In the climax, Neo becomes all-powerful within the Matrix once he awakens and sees it as a computer-generated dream world.

No wonder a block buster science fiction movie was for the first time dealing with the subject of Reality, Man and Destiny ...... which has been dealt with for ages by philosophers and thinkers all over the world.

Metropolis

German Director Fritz Lang's film Metropolis (1926) stands out as a science fiction classic film of the silent era considered one of the first full length science fiction film which still inspires awe as far as its cinematic techniques are concerned.

Filmed in 1925, at a cost of approximately five million Reichsmarks, made it the most expensive film ever released up to that point. The scale of the movie was huge which can be gauged from the fact that it had over 38,000 extras.

The film projects a scenario of 2026, the future city – Metropolis – with shiny sky scrappers piercing the sky in the clouds and the rich people enjoying life in the lap of wealth and luxury

In sharp contrast lie the vast majority of people in darkness below the earth to work in factories to provide the much needed power to run the city of Metropolis. They have been denuded of all humanity and work as servile slaves like human robots an extension of the machines.

This kind of sharp social division with high industrial technological advancement and prosperity for few of the elite on the one hand and on the other hand deprivation, squalor, poverty and back breaking labor for the large masses of ordinary people to barely survive, was presented with such impact on the celluloid perhaps for the first time that Metropolis has remained an iconic film influencing large number of film makers in the years to come.

The film was written by Fritz Lang and his wife Thea von Harbou, and starred Brigitte Helm, Gustav Fröhlich, Alfred Abel and Rudolf Klein-Rogge.

Almost all special effect techniques available at that point of time like matte painting to create large scale realistic backgrounds , rear projections where one had to show a moving background, multiple exposure shots to show rising and falling rings on the robot, animation techniques to show airplanes flying above the city, were used by Lang.

He used miniature models in an extremely skilled way taking advantage of perspective and camera angles to to blow up and create a believable true to life life modern futuristic city with towering sky scrapers. Futurist Italian architect, Antonio Sant'Elia influenced the architecture and design of Metropolis as also many other science fiction cities, infilms like Blade Runner, The Fifth Element, Brazil.

Lang used a new filming technique known as Schufftan process to insert characters in the required miniature model of the sets using mirrors, much before the advent of the blue screen technology.

The importance of this film's legacy can be grasped in the fact that the restored version from 2001 is part of UNESCO's list of Memory of the World. It was the first feature film ever to be taken up on that list.

Metropolis is regarded as one of the most influential films in the history of science fiction films, because of its special effects and the vision of city of the future.

Cameraman Günther Rittau, special effects designer Eugen Schüfftan, and art director Erich Kettelhut worked cohesively to create the design and look of the city of Metropolis which remains as a landmark in the history of cinema.

Interestingly Metropolis visualized a modern city which was something unseen in Germany at that point in time. The dynamic fast moving urban city shown in the film was more like New York in USA and Lang too was inspired by New York.

In an interview, Lang reported that "the film was born from my first sight of the skyscrapers in New York in October 1924".

"I looked into the streets – the glaring lights and the tall buildings – and there I conceived Metropolis."

The design and architecture of buildings and cityscape of Metropolis are supposed to have been also influenced by modern art movements and styles like Art Deco, Dada and German Expressionism. The Tower of Babel, headquarters of Joh Fredersen the master of the city was inspired by Pieter Brueghel the Elder's 1563 painting.

Adolf Hitler is supposed to have "loved" Metropolis and the Nazis would later embrace the film's imagery in their own visual iconography, from the Reichsadler eagle (Roman eagle standard, used by the Holy Roman Emperors), to the use of massive lines and architecture employed for Leni Riefenstahl's 1935 propaganda film 'Triumph of the Will'.

The young man Freder, protagonist of the story is the son of the Master of Metropolis. One day he sees beautiful Maria coming out from underground with children . She is the spiritual leader of the workers, and she has brought the workers' children to the surface to meet their comrades above but guards sent them back, however Freder follows them into their world which he never knew existed and soon is shocked to discover their hard brutish and agonizingly monotonous and mechanical life of drudgery and slavery.

The men have become completely mechanical like robots having lost their human free spirit. In a powerful visualization depicted by Lang, Freder has a hallucination and imagines one of the giant machines as the ancient god Moloch devouring columns of human robots. Rows and rows of workers march like zombies to their own death in the gaping wide open mouth of Moloch.

Existentialist philosopher Martin Heidegger who warned about grave negative impacts of technology is supposed may have been greatly influenced by the film.

Freder on his return confronts his father, the Master of Metropolis and also warns of possible revolt. Prepared for any such eventuality his father has got constructed a Robot by his scientist Rotwang and to control the workers gets the robot to have a face identical to Maria.

Freder sees Maria preaching messages of hope and social justice to the poor working classes, with a prophecy about the coming of a Messiah who can bring the working and ruling classes together. Freder wishes to fit that role motivated more by his love for Maria.

The false Maria robot's attempt to sabotage the workers' rebellion by instigating large scale violence and destruction and putting the lives of children at danger by flooding, backfires when Maria and Freder save them. Workers coming to know that Maria had put children at risk burn her as a witch in the stake and while she is in flames it is revealed that she is a robot.

Freder saves the real Maria from Rotwang and eventually kills him in a fight. Freder says he is the heart or "The mediator between the brain and the hands".

At the end Frederson too realizes that between the 'head' (the Masters of Metropolis) and the 'hands' (the workers), there must be a mediator – the heart.

Basically what the film has been able to achieve is to be able to portray in cinematic language the sharp class differences and divisions in a modern capitalist exploitative society full of decadence . The inherent propensity of dehumanization is due to the greater reliance on machines and to view workers manning the machines not as humans but as just an extension of machines.

So the main question being raised is that can man be treated like a machine ? Are men machines ? What is the difference between Man and Machine?

The "glossy" life style, the pleasure gardens, hundreds of storied high skyscrapers , the fast growth, and progress of a city like Metropolis is based on the large masses of working class toiling day and night to run the well oiled machinery of the city. The rich who are corrupt and greedy hold the reins of power and capital and exploit the poor to see that their factories and city keeps running smoothly while they keep enjoying pleasure and profit while workers continue to live a slavish and dehumanized life.

The typical scenario to suit a Marxist ideology where workers eventually unite to revolt and throw out the capitalist class, but Metropolis does not seem to go for this straight forward simple communist solution, particularly in Hitler's Germany where the Nazi Party message was "pro-worker yet anti-communist".

Metropolis goes in for religious symbolism and messianic role for Freder who has been shown in trappings of Christ, the film concludes with a kind of idealistic hope of forces of the workers and capitalists joining together, the hand (workers) with the head (capitalists) and Fedrer (heart)serving as the mediator.

It is the heart or emotions and feelings which differentiates Man from Machine. The stress therefore was on "compassion", "conciliation"and "co operation" between both the classes to bring about development and progress of the state.

This may have been an ideal of Nazi Germany and therefore Hitler loved the film , but was never put in practice.

As visual medium Fritz Lang has done a remarkable job in creating in Metropolis long lasting powerful images to highlight economic progress and development as also vivid images of life of slavery and dehumanization of workers, inspiring generations of film makers after him.

Planet of the Apes

Planet of the Apes was a huge success when it was released in 1968 stunning audiences depicting a complete role reversal of man and beast, portraying a world where dominant Apes ruled over servile men with strong undertone of socio-political satire of the US's racial relations in the 1960s .

The symbolism of the last shot is truly memorable with the hero Charlton Heston finding his worst fears coming true - the planet the Apes have taken over is none other than mother earth. He has somehow returned to a transformed America (which they left over 2006 years ago) with the Statue of Liberty buried deep underground with only portion of its bust and head above the earth. As the shocking truth dawns on him that perhaps humanity has destroyed itself in a nuclear holocaust, he falls on his knees, tearful and frustrated banging his fists on the ground.

Human civilization has been annhilated by the self destructive nature of Man and he cries in despair "you maniacs! you blew it up! god damn you!"

The movie ends with this shot considered to be one of the most memorable endings in the history of cinema.

Late 1960s was a period of anxiety and fear with threat of nuclear war very much alive in the period of Cold War particularly among Americans.

Planet of the Apes impact with its completely different take on the possible consequence of a nuclear war was greatly appreciated by the audiences of that day.

A little hint is given in Taylor's last transmission to earth about these anxieties where he says,"Does man, that marvel of the universe, that glorious paradox that sent me to the stars, still make war with his brother?"

Not many people know that Planet of the Apes was based on the novel of French writer Pierre Boulle called La Planète des singes. Boulle was more famous for another novel which was as The Bridge over the River Kwai (1952),which was made into a classic movie by David Lean.That film won seven Oscars, including Best Adapted Screenplay. Planet of the Apes got only one Honorary award at Oscars for outstanding make up which went to John Chambers.

Interestingly the novel had shown a much more technologically advanced society of Apes but the film decided to show them in a lot more primitive form.

Before even they land on the unknown planet a fatality occurs in the spaceship , the lone woman astronaut's suspended animation chamber has failed, resulting in her death.

The survivors - Taylor, Landon, and Dodge, find that the planet they land on is a habitable world and start searching for food and water for their survival.

They find that the world is inhabited by mute humans in a pre civilized state living like animals and being treated as such by the Apes who are the dominant species.

Apes are virtually like humans with the ability to think, walk and talk and have built a civilization almost on the lines of humans on earth with minimal tolerance and scientific temper yet to be develop with superstition being more valued than science.

In this world there are three classes gorillas (workers and soldiers), chimpanzees (scientists and intellectuals) and orangutans (governors and politicians).

Humans have intruded into the crops cultivated by the apes. Gorilla soldiers with modern firearms on horses come hunting for humans, as humans run scared in desperation to save their lives like rats in a standing crop.

They kill many of the humans, and capture them for subsequent experimentation by the chimpanzees.

Dodge is killed (whose body is later found to be stuffed in the Natural History Museum later by Taylor like we keep stuffed lions and tigers and other animals). Taylor and Landon are wounded and taken as prisoners. Landon, is subjected to a lobotomy, while Heston, suffering a neck wound which makes him mute for sometime and is forced into a cage like other homo sapiens for future vivisection.Wounded in the throat, Taylor cannot talk, but he does manage to steal a pencil and notepad to communicate with apes like Dr. Zira, an animal psychologist, and Dr. Cornelius, her fiancé and an archaeologist. The apes are shocked to find he can write. Taylor escapes, but is captured and in a dramatic scene arrogantly shouts "Take your stinking paws off me, you damn dirty ape!". The apes are astonished by a talking human.

Through the humanitarian gesture of a chimpanzee scientists Heston is temporarily spared and given an human female, Linda Harrison, to be a partner, whom he later calls Nova.

Cornelius and Zira argue on the unique man's behalf, the lead scientist, Dr. Zaius (Maurice Evans) who is interestingly also the Defender of Faith believes he should be executed and Taylor's fate is sealed. But, before he can be killed, Zira's nephew Lucius, Zira and Cornelius help him escape and take Taylor and Nova into the "Forbidden Zone".

Forbidden Zone is out of bounds for the apes for centuries as their origins are supposed to lie there. Cornelius and Zira want to go there to be able to gather proof of a non-simian civilization before the Apes (which Cornelius discovered earlier) so that all charges against them of heresy would be absolved. Before they enter the cave, Zaius intercepts them with his soldiers but Taylor threatens to shoot Zaius, who agrees to enter the cave to disprove their theories.

Inside, Cornelius displays remnants of a technologically advanced human society pre-dating simian history. Taylor identifies artifacts such as dentures, eyeglasses, a heart valve, and to the apes' astonishment, a child's talking doll. At this point Zaiusreveals that he knew about the ancient human civilization but warns Taylor that he may not like the answers.

Taylor and Nova are allowed to leave but Dr. Zaius orders the cave to be sealed to destroy the revolutionary evidence which has the potential overturn the entire ancient history of the Apes and their evolution, as any hidebound orthodox authoritarian would do. Not only that he does not allow Zira, Cornelius, and Lucius to go free with the fresh enlightened knowledge they have but charges them with heresy. With their evidence lost their fate is sealed.

Taylor and Nova leave on horseback and soon discover the top portion of the Statue of Liberty jutting out of the sandy beach . Taylor has landed on the birthplace of his own race the earth . He sinks to the ground on his knees in anguish knowing the fate of human kind and lashes out at them, "god damn you!"

The nuclear devastation seems to have brought human kind to the position it is in, possibly with their inability to have been able to keep pace with their simian counterpart who outstripped them.

The film really puts up a mirror in front of today's society and shows how religion and science are both manipulated by authoritarian powers to keep control in their own hands and maintain status quo.

In a satirical representation it depicts Dr.Zaius as both Minister of Science and Defender of Faith. He oversees religion and scientific development in the Planet of the Apes, thus hiding scientific truths or evidence regarding early human civilization in the Forbidden Zone so that the religious narrative is not falsified in anyway. He stops scientific advancements when they do not adhere to their religious history. So scientific freedom and exploration can continue but in a limited fashion.

The Planet of the Apes showing that results of war can be a virtual wiping out of the civilization and the species is also an anti war movie particularly at time when America was witnessing widespread protests against the Vietnam war and America's involvement in it.

By reversal of roles and making the protagonist kind of a laboratory guinea pig a clear message is also being sent against animal experimentation and cruelty to animals. In the sense the film does through its satire try to shake human conscience regarding its treatment of animals and for that sense all living things.

The treatment of the humans by the apes mirrors how humans treated animals in the 1960's, how they were used for scientific experiments. The sympathy that the film evokes for the humans makes the audience think about how animals in the real world are treated in the name of science. When Zira, a chimpanzee scientist, discovers that Taylor, one of the human astronauts, is in fact intelligent she believes it is an anomaly. She is interested in studying Taylor, because she recognizes him as an intelligent creature.

Even though she can recognize that he is intelligent , she is still willing to use him for her scientific research.

Besides having a novel concept for a film, the film really stands out with its outstanding makeup and costume which are extremely natural .They don't look like men wearing monkey suits as is seen in many earlier ape based films.

With Heston in the lead are veterans like Kim Hunter as Zira, Maurice Evansas Dr. Zaius, James Whitmore as the President of the Assembly, and newcomer Linda Harrison as Nova .

While the make-ups realistically simulate ape features with the creases at the correct places allowing full mobility and animation. Individual quirks and signs are painstakingly adhered to.

The dramatic music throughout the film by Jerry Goldsmith's score employs unusual instrumentation inspired from modernist composers such as Bartok and Stravinsky.

Some of the plethora of musical instruments he uses to create the surrealistic soundscape of a primal and primitive world ruled by Apes is worthy of mention.

The musical instruments include ethnic instruments and non-traditional instruments including a xylophone, a vibra slap, a cuika,stainless steel mixing bowls, and a echoplex to create echoing effects, percussion instruments like boo-bams,drums, also bells, water drop bars, a piano, an electric harp, an electric bass clarinet, a bass slide whistle, a Shofar, and a ten foot long Tibetan horn.

A Reviewer has said that music at the beginning is one of the "greatest openings in film".......

Low register piano and percussion lead to a resounding Shofar blast as the titles come on the screen.

John Chambers (a talented make-up artist responsible for Spock's pointy ears in "Star Trek") had the job to develop the simian prosthetics. Franklin J. Schaffner, directed the film.

"Planet of the Apes" was filmed near the Grand Canyon in northern Arizona, Malibu Creek State Park and Zuma Beach, Malibu.

Schaffner's movie costing $5.8 million to produce was . released February 8, 1968, and met with both critical and commercial acclaim grossing over $ 26 million.

The satire on racial relations in the movie presented a chilling allegory on the subject of racial conflict, highlighting the injustices of America's slave past, as struggle for Civil Rights continued at that time.

Taylor is captured by hunters and taken to Ape City where he is treated inhumanly by militaristic gorillas including whippings and beatings, a reminder of the height of slavery in the US.

Civil rights protesters in Birmingham, Alabama were cowed down in the 1960s with fire hose water jets and almost similar method is used on the astronauts by their captors.

The incisive socio-political theme of the movie adds to its innovative idea and along with its superb presentation it its place is assured as one of the great science fiction movies.

The Terminator

Director James Cameron's first Terminator film written by him, 'The Terminator' starring Arnold Schwarzenegger went on to re-establish Schwarzenegger's position as a leading action hero (anti-hero)and launched the film career of Cameron.

Arnold is supposed to have revealed years later that actually James Cameron 'talked him into it' by telling him that it's not going to be the 'villain' but a heroic character, as people told him that he should not play a villain as it would not be good for his career.

This cult science fiction film one of the most successful films of 1984 making $ 80 million in the box office with the film made on budget of $6.5 million. In 2008, The Terminator was selected by the Library of Congress for preservation in the National Film Registry as "culturally, historically, or aesthetically significant".

The star of the film is the six feet two inches tall 200 pound electronically controlled robot of thousands chrome plated moving parts built and designed by Special Visual Effects head Stan Winston with sketches made by Cameron with and hand crafted.

Sculpted models of Arnold were so realistic that Arnold himself has said that he would be wrong six to seven times out ten to tell difference between the model and real Arnold. It is a time travel story but with a gripping and tense storyline of a Terminator ,a cyborg (cybernetic organism with both organic and biomechatronic body parts)travelling from 2029 to the to 1984 Los Angeles on an assassination mission.

Literally a killing machine Terminator kills four people just to acquire guns and clothes and then begins to systematically killing women , whose addresses it finds in the telephone directory.

Even though it is able to track Sarah Connor to a nightclub, a human soldier sent back in time from 2029, Kyle Reese rescues her. They steal a car and escape with the Terminator pursuing in a police car.

In the future world artificial intelligence defense network, Skynet , created by Cyberdyne Systems, has become self aware gained control with number of other machines and started nuclear war against humans. Sarah's future son is leading a powerful resistance movement against Skynet but before he succeeds a Terminator is sent back in time to kill Sarah before John is born to prevent the formation of the Resistance.

The cyborg Terminator sent to assassinate Sarah has a metal endoskeleton with only external layer of living tissue that makes it appear human. Terminator goes on a mindless destruction and killing spree including shooting a number of policemen in a police station to locate Kyle and Sarah who were apprehended by police.

In the melee Kyle and Sarah steal another car and go to a motel to hide from Terminator and slowly gather bombs. Kyle also reveals his love for Sarah since he had seen her photograph with John. He had undertaken time travel just to save her. Both fall in love with each other.

The Terminator kills Sarah's mother and eventually locates Kyle and Sarah and relentlessly pursues them despite being severely damaged by Kyle's bombs. Eventually Kyle puts a bomb into Terminator's abdomen and blows him up though he himself is also killed and Sarah is injured. Terminator's mechanical torso reactivates and pursues Sarah but she slyly draws it towards a hydraulic press and crushes it. The film ends with a pregnant Sarah traveling through Mexico buying her own photograph taken by a boy at a gas station.

When the film 'The Terminator' was released no one had seen a Cyborg of the ruthless terrifying kind that Arnold Schwarzernegger depicted in the film.

The words of Kyle "Listen, and understand. That terminator is out there. It can't be bargained with. It can't be reasoned with. It doesn't feel pity, or remorse, or fear. And it absolutely will not stop, ever, until you are dead," instills the kind of fear one can have of an entity which is pure machine and completely emotionless.

Schwarzenegger is at his menacing best fully armed with state-of-the-art weaponry and goes bullet-pumping with such speed and precision that the audience is left completely stunned and amazed to even think about the merits and demerits of the action.

One of the most iconic sequence of the film when this kind of a violence is seen is in the police shootout where the Terminator massacres an entire precinct to get to Sarah.

Though it may be right to criticize the film for glamorizing violence but then Arnold does add a lot of fiendish charm to his killer machine.

Another shocking moment is when Arnold peels back his synthetic tissue on his arm to reveal mechanical gears and wires, cruel reminder that he is a machine.The scenario built up in the film about ascendancy of machines is no more just a conjecture looking at the technological advancements taking place in the field of cybernatics, Artificial Intelligence,robotics.

Intelligent robots, especially if they can match or even surpass human capabilities are very much within the realm of possibility. If these robots become rogue than it may not only be difficult but virtually impossible to control them.

One of the highly controversial subject today is LAWS (Lethal Autonomous Weapon Systems) which can act on their own without direct human control to destroy or kill a target. These killer robots can be from insect like drones to full fledged humanoid robots.

International bodies like the UN are already seized of the matter being petitioned to develop international laws and guidelines to stop or control these futuristic machines particularly of interest to the Defence establishments of many countries. One can imagine the result of LAWS in the hands of a terrorist or criminal organization.

As a sci fi film 'The Terminator' does give us glimpse of what might happen if we have such a mechanical slaughter machine amidst us.

The whole question of machine ethics is also becoming relevant at least as an issue for discussion as greater and greater human-machine interface is becoming common.

Non human machines taking over humanity is an extremely scary scenario primarily because they do not have emtions and feelings nor can understand emotions and feelings.

The higher thinking processes, motivations which drive humans can never be there in machines which are manufactured to be able to act strictly underfixed commands.

Tron

The 1982 movie Tron produced by Walt Disney Productions was a path breaking movie and cannot be missed by any work related to science fiction films, even thogh it had a large dose of fantasy element..

It was perhaps the first movie to bring on a single cohesive format animation, live action and computer graphics together in a popular film. Interestingly the film was not acceptable at that point of time for Academy Awards, as it used computers.

The film in a way pioneered the use of Computer Graphic Imagery for the film was done by four foremost computer graphic companies at that point of time, Information International Inc. (Triple-I) and Robert Abel & Associates of Los Angeles, Mathematic Applications Group Inc. (MAGI) and Digital Effects of New York.

The 15 minutes of CGI and the over 50 minutes of backlit animation drove the cost of the movie to a whopping $20 million.

 Actually Tron attempts to brings to life the world inside the computer - the digital world . What will it be like inside an electronic world ? What kind of creatures can live in that world of energy and electricity ?

The Tron is not all fantasy because of the development in the field of quantum computer and quantum teleportation shows that theoretically it is now physically conceivable to actually take a particle in the real world and teleport it into a quantum computer, at least the particle information can be send into the computer.

Stars of Tron include Jeff Bridges, David Warner, Bruce Boxleitner, Cindy Morgan and Barnard Hughes.

The story is innovative and gripping, a Programmer Kevin Flynn (Jeff Bridges) has been cheated. His former colleague at ENCOM Corporation Ed Dillinger (David Warner) has stolen the ideas for several popular arcade games that Flynn had written.

Flynn gets inside the corporate house and looks in the computer system. But Dillinger's powerful artificial intelligence, the Master Control Program (MCP) is malevolent and self aware .

After helping develop the Master Control Program (MCP) for a set of video games he'd developed, Kevin Flynn is shocked to discover that the program seems not only to have gone rogue, but developed its own allies, but it eventually gathers enough power that it begins blackmailing Dillinger in its pursuit of domination.

It has been stealing other computer programs across the world . Understanding the intention of Flynn , it uses an experimental laser to "digitize" Flynn, turning him into a computer file under its command. All of the programs in the System appear to be humans, with the faces of their "users" and glowing lines on their unisex costumes.

To make matters worse SARK, a control program (whose user is Dillinger) is tasked with eliminating Flynn. The MCP wants Flynn to be run through a series of game programs, where he will be deleted upon being defeated. Flynn, a pretty good video game player, survives his first challenge. He is imprisoned with TRON, a monitoring program but they escape on "light-cycles." Flynn as a programmer has powers in this electronic computer and is able to revive programs, rebuild broken vehicles.

Flynn and TRON assault the MCP and ultimately defeat it and MCP's various security measures are not there. Flynn is rematerialized, and TRON uncovers Dillinger's deceit. Dillinger is unseated from his high position, and Flynn takes his place.

The film Tron wasdirected by Steven Lisberger's first directorial venture which he wrote and developed with producer Donald Kushner.

Tron is the first movie to use computer graphics imagery extensively , the entire world in which the characters in Tron live and move about is not the real world but is created by CGI,the landscape and vehicles are all computer generated.

Even though computer graphics imagery had been introduced in films like Westworld, Star Trek, Looker, the first film to make heavy use of CGI is Tron.

Tron is set in two worlds: the real world, where a vast computer system of a communications conglomerate is controlled by a single program; and the electronic world, whose electric-and-light beings want to overthrow the program which controls their lives.

Richard Taylor, who headed the Entertainment Technology Group at Information International Inc.co-supervised the special effects on Tron. He oversaw the design and programming of the film's computer animation.

There are over two hundred scenes that utilize computer-generated backgrounds. Much of the remaining effects in the film were backlit optical effects. According to Taylor to join seamlessly computer simulation and live action photography was one of the major challenges of the film. Taylor has pointed out that they were shooting people in black and white costumes on sets that were black, and matting those people into computer simulated worlds and at times putting computer simulated images into graphically created scenes. They had to homogenize all that so that it did not seem to be done at different places but all at one place.

Creating the first film heavily based on computers they had also to make the computer professionals aware of the needs of the film makers because the computer professionals did not approach "creation of visual imagery the same way that a film-maker will".

Similarly, the film maker also came without any knowledge of computers but what they wanted on the film.

Director Steven Liserberger was fascinated with the image of graphic movement through computer generated environment. With video games becoming popular "I started talking to the computer people, it seemed that I now had the characters I could put into that computer environment," Liserberger has said.

In the mid-seventies, Steven Lisberger was operating a studio in Boston that was producing animation for commercials he began exploring the possibilities of using computer-generated imagery for a story about characters that lived inside a video game.

Using conventional animation, Lisberger produced a animated logo for his studio, which was licensed as advertising to several radio stations around the country. This was the first appearance of a character that he called Tron.

In 1977, Lisberger partnered producer Donald Kushner to produce an animated spoof of the Olympic Games entitled *"Animalympics."* In June, 1980, they approached The Walt Disney Studios with a detailed proposal for the project. Their ideas were enthusiastically received, and pre-production began shortly thereafter.

According to Kushner they used computer simulation, used backlit techniques, and conventional live action but the difficult part was to make them all look like a part of a cohesive work, which was eventually achieved.

According to Stephen Lisberger those days there was no movement and computers could only generate individual frames. There was no way to digitally put them on film so you actually set up a motion picture camera in front of a computer screen and you filmed it frame by frame. Some of the frames took hours to generate. It was created on a computer that only had 2MB of memory and a disc with only 330MB of storage

Tron was not nominated for Oscars, Lisberger once commented on it "Tron was so unique that there really wasn't anything to compare it to. So, in a way, they solved the problem by just ignoring it."

For example, the lightcycles and other vehicles were animated, not computer-generated using 75,000 frames of "special animation cels" called Kodaliths.

Associate producer Harrison Ellenshaw has stated, "We were not trying to reinvent the wheel. We were trying to make a whole different wheel that was not necessarily round... All movies until Tron were set in an established genre. Tron created a new one."

The laser battles were filmed at Lawrence Livermore Laboratory in its linear accelerator using fluorescent lights to create strobe effects.

Tron also used light sources underneath the characters' costumes, which were white with black lines, to help create the glowing effect. The actors were also filmed against a black background, not green or blue like today.

CHAPTER 7
INDIAN SCIENCE FICTION FILMS

Indians are highly emotional people with strong family ties ,they may be very credulous and superstitious but are extremely imaginative and expressive in their variety of cultural traditions like literature, song, dance, drama and music. So when Indians make even science fiction films it has to become a part of the existing thematic storylines involving emotional family drama, love stories, rivalries and revenge intrigues with lots of music, song and dance sequences.

Renowned Science Fiction Hollywood director Christopher Nolan has also said that Hindi films with their emotion, drama and action involve the audience more sensorially, one of the fundamental reasons due to which we enjoy Hindi cinema while Hollywood had lost some of that essence. The Indian science fiction film directors understood that their film has to fit into the time tested formulas of Indian films to succeed in the box office.

India has a glorious past in the field of science. Even internationally renowned scientist Albert Einstein has said, "We owe a lot to the Indians, who taught us how to count, without which no worthwhile scientific discovery could have been made."

But despite Indians having knowledge of speed of light described in Rg Veda, Bhaskaracharya having spoken about gravity more than 500 years before Newton, Aryabhatta describing distances of planets from the sun in the 5th century, and Indian scientists even today making a name for themselves all over the world, large masses of Indians still lack what we calla scientific temper. For the Indian science fiction film makers the one area where they take poetic license is the "science" aspect of the story. Except for a few rare films, scientific notions and ideas are perfunctorily handled. Anyone going about with a tousled crop of white hair like Einstein is dubbed as a "scientist" or a "doctor".

The filmmakers serve the public misconception about scientist and doctors by presenting them as "absent minded "professor or "mad" scientist, rather than trying to show thema s a serious level headed person like any scientist or doctor in real life.

Similarly laboratory has always meant in Indian films just a few test tubes, funnels and flasks, or some boxes with flashing lights and electrical tubes.

There is still a lot to be demanded as far as realism is concerned in projection of scientific experiments and scientific ideas in Indian films.

In the run of the mill sci fi Indian films, you have the scientist as the character who possesses the "secret" formula or "potion" which can cause either some great disaster or bring about a beneficial transformation. The scientific rationale or details of the formula or the potion are never spelled out and the viewer is usually kept in the dark.

Typically the film's villain who maybe the enemy of the Indian nation or a gangster wanting make profit by selling the formula kidnaps the scientist or steals it from him. The film's hero then goes about rescuing the scientist and getting the formula back.

On a slightly different track a science fiction film may also have a machine which can transport you in time, sometimes it may not be a machine but just a wrist-watch. Again there is not much elaboration of the science behind it.

Even though Indian film industry has made a number of sci fi films, they are few and far between though the trend seems to be changing lately.

Large number of Indian films are slated as or publicized as science fiction films to appeal to what the film makers feel is the growing audience for this genre. Even though many of these films are profitable at the box office most of these do not fall strictly under the category of science fiction films.

At times the plot of the story just has an extremely tenuous or extraneous connection with science and that too seems to have been made with great difficulty to be able to call the film a "science fiction film" while the entire film turns out to be a full scale suspense, horror, comedy movie. Sometimes the science fiction film just peters out into complete fantasy and magic full of unscientific and bizarre happenings leaving the audience looking for good sci fi movie completely at a loss.

These kind of films succeed primarily because of the willing suspension of disbelief by the audience who are ready to keep at bay their logical and critical faculties to just immerse themselves and enjoy the show.

Adhering to science is difficult and many top Hollywood science fiction films too are riddled with scientific flaws and loopholes.

Creative license in a science fiction film should not mean that you can show what is blatantly irrational, illogical and unscientific. Indian films somehow lagged behind in this area.

Indian films which have tackled a scientific subject with some seriousness are extremely few.

What do insiders have to say?

Bengal Film Director Nitish Roy who has made the sci fi film Jole Jongole and several other films, is a well known Art Director and Production Designer even in Hindi cinema and has worked with top directors like Shyam Benegal, Mrinal Sen, Govind Nihlani, Rajkumar Santoshi, and international directors like Mira Nair and Gurinder Chadha.

He designed and created Roman architecture sets for renowned Oscar winning Hollywood film Gladiator. He has also given architectural input for the creation of film cities like Ramoji Film City (Hyderabad) Innovative Film city (Bengaluru), Prayag Film City (Kolkata) and several theme parks, amusement parks, museums not only in India but also other countries.

Mr. Roy attributed poor status of sci fi films in India due to a number of reasons in an interview.

Q. 1. Science fiction films continue to be very few in the country particularly in the regional languages. Bengal has a strong foundation of sci fi literature but does not reflect in the films which continue to be very few. Why? What are the constraints?

A. I really appreciate your question. Bengal, not only has a strong foundation of literature in the field of science fiction there are some truly masterpieces which can be canvassed magnificently. But reality speaks in different note. Well the main reason that I presume is lack of interest and budget restriction for making Sci-Fi. Producers doesn't mind wasting money on big names but they are conservative when it comes to making Sci-Fi films. There are a lot of evidences that can substantiate my saying. That's the reason the Directors are now days not at all interested or inclined in making Sci-Fi.

"Any sci-fi film made in India we always have a tendency to compare with that of Hollywood but one never compares the budget involved in Hollywood and Indian Sci-Fi films which is most unfortunate. That's also a reason why distributors have reservation in doing business in this particular genre."

Q2. As a film director how has been your experience in making science fiction films?

A. If you ask me about my experience in making science fiction films as a Director, it will be difficult for me to actually make you understand. It's wonderful, thrilling and above all very satisfying. But again, the indifferent attitude while releasing the film makes me sad.

Q.3 Many so called science fiction films in India are actually pure fantasy like Super Hero films? Do you think lack of interest in science or a scientific temper has something to do with this ?

A. Science Fiction films should not be confused with Super Hero Films. They are totally different. I don't think there is any lack of interest in science or a scientific temper. Basically the audience haven't been given a chance to have a look at Sci-Fi films.

Q.4 Science fiction films particularly require sophisticated set designs, hi technology whether practical Special Effects or CGI, as an Art Director do you feel India particularly regional film industries have the kind of capability to produce high grade sci fi film like that of the west ?

A. Capability is always there and can never be questioned. Hi-tech special effects or VFX needs to be infused wisely without hampering the story line. I strongly believe that we are even capable of delivering better than those produced in the West. The only hindrance is the free hand given to the Director in Western countries is more unconditional compared to the unnecessary interference that is prevalent in this part of the Globe.

Q. 5 What kind of special effects were used in JoleJongole What was your experience in making that film ?

A. We have used live animation, VFX and a 100ft long pneumatic operated crocodile in Jole Jongole. The experience was simply outstanding. It was star studded film starring Mithun Chakraborty, Ashish Vidyarthi, Tinnu Anand, Dijana Dejanovic, Biswajit Chakraborty, Mumtaz Sorcar. The story line as well as the visual effect was simply exemplary.

Q. 6. Are others in the pipeline?

A. Yes, after Jole Jongole, I have made a true Sci-fi named 'Nonoga' based on Shirshendu Mukherjee's novel 'Gondogol'. Its again a star studded film featuring Victor Banerjee, Bratya Basu, Kharaj Mukherjee, Konineeka Banerjee, Bhaskar Banerjee and others. It was supposed to be released in May 2020, but the pandemic situation forbade us to release the film. This is the first of its kind in pan India Film industry. We are thinking of releasing the same during the course of this year.

Award winning renowned production designer and art director Sabu Cyril has won National Film Award for Best Production Design four times, Filmfare Award for Best Art Direction (including south) five times. He was nominated at 42nd Saturn Awards for his work on Baahubali: The Beginning and has won international recognition for his work in Tamil language sci fi film in Enthiran .

In an interview Mr Sabu Cyril pointed out financial constraint being one of the main reasons for India not being able to make convincing sci fi films.

Q 1. Why do you think science fiction films made in India have somehow not generated sufficient amount of backing from producers and distributers, neither have many directors attempted to make science fiction films on a regular basis ?

A. Since the efforts we have made so far have not garnered a positive response at the box office, thereby, making the genre seem experimental and not necessarily viable for producers to invest in. Without an adequate budget a convincing science fiction film cannot be realistically tackled.

Q 2. Why except a very few films Indian audience which seems to be quite keen on Hollywood sci fi films so much so that many English films have been dubbed into Hindi and also regional languages, have not shown similar enthusiasm for Indian made sci fi films?

A. There are different expectations from a Hollywood film as opposed an Indian film. Hollywood films do market research to capture the imaginations of a wider, more universal, audience whereas Indian films cater specifically to an Indian audience.

Even within the country Hindi audiences view Telugu films very differently from how they view Hindi cinema. So, even if there is enthusiasm the universality and appeal is limited, with respect to Hollywood. One of the highest grossing Indian film is Bahubali 2 at about $260 Million whereas Avengers Endgame is around $2.8 Billion… So can't really compare one to the other.

Q 3.Do you think there is a lack of scientific temper or lack of understanding as far as science as a subject is concerned for the audience and also the film makers?

A. There is definitely a different framework from which audiences perceive cinema here.

Science itself is not very entertaining for most to begin with and science fiction invariably enters the realm of myth and fantasy which is already a very strong genre in our country. Concepts of flying chariots, looping and variability of time, false realities and objects/animals showing human/magical characteristics are subject areas that have been presented very differently in our books of mythology such as the Ramayana and Mahabharata.

Q 4. A pure science fiction film is one where we see the affect or impact of a scientific discovery, invention or principle on a person or society, whether at present or at a future time or even in the past; it can be anywhere on earth or at any other place in the universe. Creative speculation is allowed but the sci-fi should not become a complete fantasy or horror. It must remain true to its genre by basing itself firmly on science and logic. In India many films touted as science fiction turn out to be pure fantasy where magic rules rather than any logic or science. Do you feel we need to make more well grounded pure science fiction films and make these distinctions clear (Superman is a fantasy but Iron Man is not because his gadgets are based on technology)?

A. Yes, definitely. There is a difference between Comic Book and Science Fiction films. Dark Knight belongs to the former while Westworld the latter. However, even in America the Crichton's 1973 Westworld was ahead of its time and Jonathan Nolan had to reimagine the telling of the same tale as a web-series for it to be accepted today.

Q 5. All science fiction films generally need exotic locales, even characters like Robots or Aliens or pre historic animals etc. The Production Designer/ Art Director becomes the most crucial person in a science fiction film. Do you agree and can you explain the vital role of the Production Designer/ Art Director particularly with regard to sci fi films which you have been involved with like Ra One and Erithran?

A. Yes, it is the sole responsibility of the Production Designer to convincingly create the world the writers and directors have envisaged. Be it period, futuristic or sci-fi films legitimacy can only be achieved based on facts, physics and real records. Choosing the correct Production Designer, therefore, becomes an extremely crucial decision as the believability of the visuals can make or break the film. For instance, Enithran was supposed to be showcasing present time with futuristic elements, therefore, it had to be innovative yet believable in that regard.

Q 6. Traditional special effects and CGI both have important role to play in films ? What is India's standing in the world as far as Special Effects and CGI is concerned ? These technologies are particularly important for science fiction films? Are we well equipped technologically to come out with world class sci fi films comparable to Hollywood today?

A. India stands shoulder to shoulder with most of the big VFX houses in the world. We have the technology and the technicians. Films from all over the world come to India for the VFX. Life of Pi won the Oscar for best VFX (2013) where the VFX was done by Rhythm and Hues which at the time was a subsidiary of Prana Studios (an Indian studio) Ex Machina won the Oscar for best VFX (2015) and the VFX was done by Double Negative which is a subsidiary of Prime Focus (an Indian studio), '1917' won the Oscar in 2020 for best VFX, and was produced by Reliance Entertainment (an Indian studio).

Q7. Do let us know of your personal experience, challenges, achievements in films like Bahubali,Om Shanti Om, Enithran,Ra One etc

A. Every film I take it up is because it is challenging. Every film poses a new but solvable problem and it is my task to find a unique solution to these problems. Crisis management and creative solutions are the production designers job after all.

Q8. Do give us some idea of the history and development of Production Design/Art Direction and special effects technology in India over the years. How do you look at future of Production Design/Art Direction particularly in the context of slowly growing interest in sci fi ?

A. As time and technology progresses so does the production design and look of a film, as it needs to be acceptable and palatable for the audiences. Production Design and Art Direction has been around since the time of plays and theatre and sets and costumes have constantly had to adapt and resonate with the time. As there is a new appetite for sci-fi the only ally the Art Department has are books on science, technology and design so as to convincingly create a backdrop of a world that can be accepted by the audiences.

With Roy all set to release his next Bengali sci-fi directorial venture 'Nonoga' and Cyril ready to handle the "new appetite for sci fi" it seems Indian science fiction films may also have a bright future.

In today's Indian film scenario with the mandatory hero-heroine love story sequence, a couple of dance numbers and the usual parting of ways with a sad song, the scientist is introduced just to fulfill any plot requirement for suspense and drama.

But science fiction genre which basically means having a film which draws its inspiration from a scientific idea or development and extrapolating it to see how it affects an individual or society and brings about sweeping changes, either now or in some other time or place, has yet not come in a bigway in Indian films.

In the west an attempt is made even by the academic and scientific community to try and find possible scientific explanations and reasons even for super hero stories or films on the face of it seem to be totally unscientific or fantasy.

In India we do not see the scientific or academic community take interest in the world of science fiction films nor we have serious debates on issues raised by Indian science fiction films in the academic circles.

Perhaps this would happen once we have science fiction films being made in the country in a big way. The first department of film studies in the country in a University started in the prestigious Jadavpur University in 1993.

Head of the Department , Department of Film Studies , Jadavpur University answered a number of questions regarding the status of Science Fiction film in India.

Q. Why have science fiction films not taken off in India like in other parts of the world despite the fact that science fiction films from Hollywood have flooded the market and done quite well here ?

A. This is difficult to answer. In Indian cinema, the notion of genre is rather weak. In the 1960s, we found characteristics of many genres (crime thriller, comedy, domestic melodrama etc) within a single film rather than a film being of a particular genre. Generic tendencies will be more prominent after the 1990s.

This is probably because during this period, hindi cinema was more pronouncedly addressed to the middle-class. Thus, whenever sci-fi features were available earlier it was subsumed under the fantasy genre rather being more generically science fiction.

Q. In Bengal too we do not see many science fiction films even though it has a very rich tradition of science fiction literature ?

A. One of the greatest unfinished film was Satyajit Ray's *Alien* which was supposed to be a bilingual (English-Bengali) film to be produced by Columbia Pictures. It was generally considered that a sci-fi film is expensive which is beyond what a regional film industry can afford. In recent years, many attempts with sci-fi strands have been attempted.

Q. Indian producers and directors say that financial constraints with heavy requirement of Special Effects in science fiction films is one reason why many do not want to risk getting into this genre ? What is the solution?

A. I don't think this is an excuse in recent decades. If you consider a film like *Bahubaali*, Indian cinema has proved to be capable of producing special effects in global standards.

Q. Do you think science fiction films play an essential role in the development of scientific temper in people particularly students?

A. Of course. I personally remember that sci-fi stories enhanced and triggered imagination and interest in science in my childhood.

Q. There is often a criticism that Indian science fiction films do not take the science aspect of the film seriously taking liberties with it , while the typical emotional family melodrama, music and romance takes a pre eminent place. Do you think this is a valid criticism ? Is it done keeping the box office in mind as a serious sci fi film may not be acceptable to Indian audiences?

A. This has been generally the case. The melodramatic paradigm overwhelms the other aspects of a film in the case. This might be explained as bowing down to the box-office, but we should understand that the popular industry largely functions within the melodramatic paradigm. There is many a times not an option.

Q. Why do you think science fiction films made in India have somehow not generated sufficient amount of backing from producers and distributers , neither have many directors attempted to make science fiction films on a regular basis ?

A. The producers and distributors might find such a project not to be financially viable. But I think Indian directors function largely within the (social) melodramatic and (social) realist paradigm and there is no generic system to attest that sci-fi films have been regularly attempted (unlike, for example, gangster genre which is largely successful) to inspire him.

Q. Why except a very few films Indian audience which seems to be quite keen on Hollywood sci fi films so much that many English films have been dubbed into Hindi and also regional languages, have not shown similar enthusiasm for Indian made sci fi films?

A. I think Indian audience has neatly compartmentalized approach to Indian and Hollywood films. What they expect and enjoy in Indian films is not what they do in Hollywood films.

Q. Do you think there is a lack of scientific temper or lack of understanding as far as science as a subject is concerned for the audience and also the film makers ?

A. Yes, this might be the case.

Q. A pure science fiction film is one where we see the affect or impact of a scientific discovery, invention or principle on a person or society, whether at present or at a future time or even in the past; it can be anywhere on earth or at any other place in the universe?

A. In critical terminology this is called 'Hard Science Fiction'. 'Soft Science Fiction' many a times deals not with science and technology as a discipline, but with the social, psychological, ethical implications of changes in science and technology.

Q. Creative speculation is allowed but the sci-fi film don't you think should not become a complete fantasy or horror. It must remain true to its genre by basing itself firmly on science and logic?

In recent decades of speculative fiction where mixing and bending of genres are in vogue, so mixing sci-fi, fantasy and horror in a cultural text is something you often encounter; but yes, science fiction should be firmly based on science and logic.

Q. In India many films touted as science fiction turn out to be pure fantasy where magic rules rather than any logic or science. Do you feel we need to make more well grounded pure science fiction films and make these distinctions clear ? (Superman is a fantasy but Iron Man is not because his gadgets are based on technology)

A. Yes. To keep the foundations of sci fi well defined and separated from fantasy and other speculative fiction, the distictions should remain clear.

Q. Has sci fi become a part of the Film Studies in courses teaching films in India ?

A. I do not think this is much studied because the genre is such a minority, actually has never taken off properly, in Indian cinema. For a genre to be studied, one needs a steady and regular output rather sporadic exceptional instances.

Anindya Sengupta is Assistant Professor at the Department of Film Studies, Jadavpur University, and the current Head of the Department. He has recently submitted his doctoral thesis on Auteur criticism and Satyajit Ray. He has also written a science-fiction novel titled Aparthibo.

Hindi Language Films

One of the major science fiction film in recent time is Mission Mangal which is based on perhaps for the first time in Indian films on hard core "real" science. It is creditable that Bollywood has really matured and has not tried to make caricature of scientists but made an attempt to show the real life of scientists.

However the only problem with this kind of a films that it tends to become more like a documentary film. However Mission Mangal with taut script fine acting makes for a riveting and interesting movie.

True to the Indian ethos it has adhered to emotional drama in this movie too. Actually science fiction films in the west too have emotional drama and at times a sense of patriotism but nothing can beat the emotional melodrama pervading Indian films even if it is a sci fi film.

The movie Mission Mangal was made in close co operation with scientists from ISRO, therefore as far as the science part is concerned it tends to be factual and definitely goes to enhance the knowledge of the ordinary viewer regarding space flights and the innumerable problems encountered by scientists.

Mission Mangal was released in 2019 directed by Jagan Shakti and jointly produced by Cape of Good Films, Hope Productions, Fox Star Studios, Aruna Bhatia, and Anil Naidu. The film stars Akshay Kumar, Vidya Balan, Sonakshi Sinha,Taapsee Pannu, Nithya Menen, Kirti

Kulhari, Sharman Joshi, H.G. Dattatreya and Vikram Gokhale.

The gripping movie deals with India's first interplanetary expedition to Mars by ISRO scientists. A patriotic movie it brings alive the life of the scientists and technicians involved in surmounting all kinds of challenges on earth and in space to be able to achieve their goal in their first attempt.

Akshay Kumar as a ISRO scientist takes the rap for a failed mission on himself but then takes up the challenge which no one thinks is possible-a space mission to Mars and succeeds in a time frame of 24 months.

With large number women playing the role of scientists it is creditable that the film perhaps one of the few films is openly acknowledging the vital role of women in Indian space programme.

One of the earliest science fiction film in hindi was "Shikari" released in 1963 directed by Mohammed Hussain and inspired by King Kong with stars like Ajit, Ragini, Helen, Madan Puri, K NSingh, TunTun, among others.

Interestingly the film was more known for its melodious songs like "O, Tumko Piya,Dil Diya", "Mangi Hain Duaayen", "Ye Rangeen Mehfil" and "Agar Main Poochhoon" which are popular even today.

Mr. Kapoor and Jagdish partners in a circus go on an expedition to jungles of Malay to capture Otango a large sized ape along with their friend a scientist Professor Sharma and his daughter. In the jungle they meet an eccentric scientist Dr. Cyclops who is experimenting how to turn humans into gorillas. The expedition members save themselves from Dr Cyclops and after a lot of adventures, they are able to get out of the jungle and Oatango falls in lava and dies.

There are number of films made in Bollywood which take inspiration from the Invisible Man of science fiction writer H.G. Wells and use the concept as the main story or as part of the story, whether it is Kishore Kumar's Mr. X in Bombay, Emran Hashmi's MrX, Anil Kapoor's Mr India, Vinod Mehra's Mr Elaan, Naseeruddin Shah's Chamatkar, Tusshar Kapoor's Gayab, Ambitabh Bachchan's Bhootnath.

Elaan is a 1971 thriller film directed by K. Ramanlal. The film stars Vinod Mehra, Rekha, Vinod Khanna, and Madan Puri.

Naresh Kumar Saxena lives with his widowed mom and sister, Seema. He works as a freelance photographer and journalist. One day he meets with Mala Mehta and her father, who is the editor of a newspaper. Mr. Mehta hires Naresh and assigns him to go to a remote island to investigate and exposé some illegal activities taking place there.

Naresh goes there in the company of his friend, Shyam. Unfortunately, they are caught by the island guards lodged in a cell along with two others, one a scientist and the other is Ram Singh. The scientist soon dies but confides in Naresh that he has invented an atomic ring that when inserted in someone's mouth will turn that person invisible. Naresh takes the ring and puts it in his mouth becomes invisible and escapes with gangsters after him and after a number of adventures the gangsters are finally brought to book.

Mr India is one of the most well known sci fi blockbuster with the invisible man theme in Indian films .The film was released in 1987 directed by Shekhar Kapur and starring Sridevi, Anil Kapoor and Amrish Puri.

Amrish Puri in his role as evil Army General Mogambo with his pet dialogue "Mogambo khush hua" (Mogambo is pleased) became a household phrase besides Sridevi's popular number "Hawa Hawai" becoming a nationwide hit.

Arun Verma is an orphan, a street violinist and philanthropist who rents a large, old house in which he houses ten orphaned children and takes care of them with the help of caretaker Calendar. Arun is seldom able to make his ends meet, and owes many debts, so he decides to rent out the room on the first floor. Seema Soni, his first tenant, is a journalist who eventually becomes friends with everyone and Arun falls in love with her.

Arun receives a letter from a family friend, Dr. Sinha revealing that Arun's father a renowned scientist had created a device that would make its user invisible. Arun accompanied by Jugal, enters his late father's old laboratory and get hold of the device. When it is activated, it makes the wearer invisible unless red light is focused on the wearer. Seema is imprisoned by some criminals and Arun rescues her being a invisible person introducing himself to her as "Mr. India". Seema falls in love with her rescuer.

The head of the criminals Mogambo (AmrishPuri)has bombs disguised as toys, planted in public places. Tina an orphan in Arun's house , finds such a toy and dies in its blast. Arun, Seema, Calendar, and the surviving children are all captured and brought before Mogambo who tortures them to find Mr.India's true identity and the location of the invisibility device. Arun eventually admits he is Mr India but has dropped the device somewhere. All of them are imprisoned but escape soon after.

Mogambo wants to cause wide scale destruction in India by launching four Inter Continental Ballastic Missiles. Arun reaches the spot and deactivates the launch, the missiles detonate there itself destroying Mogambo's fortress and killing Mogambo in the blast. Mr India is a cult movie and was the highest grossing film of 1987. On the centenary of Indian Cinema, Mr. India was included amongst one of the 100 Greatest Indian Films of All Time.

Bollywood's extra terrestrial movie clearly seems to have been inspired by Hollywood's ET is 'Koi... Mil Gaya' which was released in 2003, even though its director Rakesh Roshan has stated that Koi...Mil Gaya is "not an

Indian E.T." and that he was more inspired by Satyajit Ray's story Alien, where a mentally challenged boy comes in contact with an alien.

The film starring Hrithik Roshan and Preity Zinta, Rekha, received the National Film Award for Best Film on Other Social Issues. It was screened at the Jerusalem Film Festival and the Nat Film Festival in Denmark. The film won Bes tFilm, Best Director (Rakesh Roshan), and Best Actor (Hrithik Roshan) in various major Bollywood award ceremonies, including the Filmfare Awards and Screen Awards.

In Koi Mil Gaya, scientist Sanjay Mehra (Rakesh Roshan) living in Canada has spent his whole life trying to contact aliens. He sends variations of the syllable "Om" into space through his computer hoping to contact Aliens. One evening he receives a reply and decides to go to the Research and Space Centre to share his findings. However, his colleagues laugh at him. Disappointed Sanjay returns home with his pregnant wife Sonia (Rekha) in the car. Suddenly while the lights and the radio begin to turn on and off a glow illuminates the sky, and a spacecraft appears overhead. In the excitement Sanjay drives off the road and is killed when the car explodes. Sonia survives the crash and returns home to India. When her son Rohit Mehra (Hrithik Roshan) is born, he is developmentally disabled due to the injury during the car accident.

She raises him in the hill station town of Kasauli where he becomes very popular despite being mentally challenged, however studies still remain unsurmountable for him leaving him many years behind others. He is also terrorized by classmate Raj and his gang. Raj is the local basketball champion and son of the district collector. Later, Nisha (PrietyZinta), a young lady and childhood friend of Raj, comes to Kasauli. Knowing Rohit wants to study computers, Nisha promises to help him study. They find Sanjay's old 'Om' computer and begin using it. Raj and his friends are antagonistic to Rohit and beat him and tell him to keep away from Nisha. Rohit goes home and prays to God to be stronger. Suddenly, Sanjay's computer starts receiving signals from the aliens, Rohit communicates with them, and the lights and electricity in the town begin to short circuit. His mother notices this and starts slapping him for using the computer. However, the lights in the city go out, and a huge spaceship appears over Kasauli. The visiting aliens leave in haste, leaving one behind.

Rohit and Nisha find and befriend the little blue alien, naming him Jadoo ("Magic") and discover his psychic abilities. Chief Inspector Khan (Mukesh Rishi) realizes one of the aliens has been left behind so decides to capture the alien, dead or alive.Jadoo discovers that Rohit is mentally challenged and uses his powers, derived from sunlight, to enhance Rohit's mind. The next morning, Rohit finds himself with a clear vision; he later solves a 10th standard mathematics problem, and becomes popular in school. Having becoming physically and mentally strong he beats up Raj's friends when they try to harass him. His mother also learns of Jadoo's existence and intends to hand him over to the authorities. However, when she learns that he cured Rohit, she embraces him in gratitude.

Rohit wins a dance competition and later he and his friends win basketball against Raj and his group due to their enhanced powers. Jadoo is spotted by police who catch him but Rohit rescues him and on persuasion of his mother plans to send him away to the safety of his home planet despite knowing that if Jadoo leaves all Rohit's powers will go too and he will revert to his earlier state. But before this police arrive at Rohit's place knock him out and take Jadoom but Rohit again saves Jadoo from police and send him off in his space craft which is summoned to take him. Rohit even though loses his powers finally gets public recognition for helping the Alien Jadoo , Nisha also tells him that her love for him has not changed despite his disability.

At the end of the movie when Raj and his friends again harass him, he is able to give them back as Jadoo returns his powers back to him permanently. There were two sequels Krrish and Krrish 3 but they were more into the fantasy and superhero category rather than science fiction.

The issue of cloning and the problems it can lead to if the clone turns out to be rogue is the subject of the hindi sci fi movie " Jaane Hoga Kya" released in 2006. Directed by Glen Barretto & Ankush Mohlait stars Aftab Shivdasani and Bipasha Basu.

Siddharth Sardesai (AftabShivdasani), a young scientist from Indian Medical Research center with help of Aditi (Bipasha Basu) daughter of a big industrialist sets up a laboratory in a mill . Siddharth first clones a mouse and then clones himself.

The whole story revolves around the clone missing and Siddharth being put behind bars for crimes committed by the clone like assaulting a girl at a Night Club and then committing a murder of a doctor. A clone is supposed to be your exact replica ? But the pertinent question which arises is , is even your exact replica of 'you'. If a crime is committed by your clone is it right to punish you, as far as society and police are concerned your clone is you.' What is the difference between 'you' and your clone ?

The film surely serves as a warning for the ethical use of cloning which is a subject mired in a lot of fear and controversy. Even though claims are made from time to time that humans have been cloned there have been no scientific evidence as yet and many religio-ethical issues have yet to be resolved .

Cloning is taken as a direct affront to the view that life is a gift from God and bringing into being a new human by cloning as opposed to normal sexual reproduction may tantamount to be an act against God's creation and Man trying to play 'God'.

The film leaves the hero in a quandary because the blame falls on him (Siddharth) who has to prove his innocence and also destroy the evil clone.

Love Story 2050 made at a cost of $12 million budget is a 2008 Indian science fiction-romance filmd irected by Harry Baweja, extensively shot in Australia.

"Love Story 2050"a time travel romance recreates a futuristic Mumbai city complete with flying cars, androids and airborne fight sequences. There is a robot teddy bear which performs different errands. The film stars Harman Baweja and Priyanka Chopra, Boman Irani and Archana Puran Singh.

Karan Malhotra is a high spirited young man, his beloved Sana is a shy girl who lives life by the rules. In a case of opposites attract they fall in love.

A scientist, Dr. Yatinder Khanna, Karan's uncle has dedicated 15 years of his life in building a time machine. Sana expresses a wish to time-travel to Mumbai in the year 2050, but she is killed in an accident before her marriage to Karan.

Karan wishes to travel forward in time to find Sana. Dr. Yatinder, Karan, and Sana's siblings, Rahul and Thea, travel forward in time and reach Mumbai in 2050. Mumbai, now has flying cars, holograms, robots, 200-story buildings. Ziesha, the reincarnation of Sana is there in the year 2050.

Ziesha is a popular singer in 2050 who does not remember her past life, but gets flashbacks of it after meeting Karan. However, after reading Sana's diary, Ziesha ultimately remembers her past life. Karan goes back to2008 in thetimemachinewithZieshaand therest of his family.

Action Replay

Bunty (Aditya Roy Kapoor) doesn't want to marry his girlfriend because he's seen his mom, Mala (Aishwarya Rai) and dad, Kishen (Akshay Kumar) spend their life fighting with each other.

Director VipulAmrutlal Shah's film Action Replay take us to mid 1970s, where we meet Akshay Kumar with long hairs and bell-bottom.

The film is inspired by the 1985 film of Robert Zemeck is Back To The Future .

Bunty's girlfriend, Tanya (Sudeepa Singh), keeps proposing to him but he always rejects. She takes Bunty to her grandfather, Anthony Gonsalves (Randhir Kapoor), a scientist who has made a Time Machine.

On their 35th anniversary, Bunty's parents, have a serious fight as Kishen's friend Kundan (Rannvijay Singh) bullies him and makes fun of him and his wife Mala supports him. The couple decide to divorce following this fight. Bunty in panic decides to use the Anthony's time machine to go back in time and make his parents fall in love.

He goes back to 1975, a time before his parents got married. Kishen as a not only weak and not good looking but lacked social skills compared to Mala who was outgoing and beautiful. Even then Mala used to make fun of Kishen.

Bunty's task is virtually cut out for him and he succeeds in making Kishen smart and eventually Mala falls in love with him Creating further complications Mala's friend Mona (Neha Dhupia) falls in love with Bunty and Kishen's parents disapprove of the Kishen's marriage with Mala. After several dramatic events finally their parents accept Kishen and Mala's relationship.

Mona proposes to Bunty but he tells her that he loves another woman. Bunty then uses the time machine to return to the future and finds his parents are in love with each other . Bunty who was reluctant to marry at the end proposes to Tanya and she accepts.

The film is a fine example of a science fiction tale of time travel fitting perfectly into a typical adventure romance family melo drama.

PK is a science fiction satire, its hero is an alien but in a departure from the usual films director Rajkumar Hirani utilizes the alien to expose the hypocritical god men, religious dogmas and superstitions prevailing inthe society with a lot of incisive humour. The film stars Aamir Khan with Anushka Sharma, Sushant Singh Rajput, Boman Irani, Saurabh Shukla and Sanjay Dutt. The alien's view of our customs and practices and his actions and comments makes a very interesting film.

The film received eight nominations at the 60th Filmfare Awards, winning two. Additionally, it won five Producers Guild Film Awards, and two Screen Awards. PK garnered the Telstra People's Choice Award at the Indian Film Festival of Melbourne. At the time of its release it emerged as the highest-grossing Indian film of all time, and ranks as the 70th highest-grossing film of 2014 worldwide. The film's final worldwide gross was Rs 854 crore.

TAMIL LANGUAGE FILMS

One of the first science fiction films of India came from Tamil Nadu in 1963 directed by A. Kasillingam with the legendry actors M.G.Ramachandran, Bhanumathi, Rajasree and M.N.Nambiar.

The name of the film was "Kalai Arasi" and it had aliens, spaceships, astronauts wearing anti-gravity boots. Seeming to be completely at ease in an interplanetary scenario it shows astronauts planning to visit earth or Boomandalam and charting out a course for it, on a planet far away from earth. A flying saucer carrying the aliens is shown moving across Jupiter and Saturn in space. The aliens are wearing metal boots and shiny clothes.

The reason for the aliens coming to "bhoomandalam" is to kidnap an artist to their own planet because they lack art and culture despite their high level of science and technology.

Obviously they have monitored different planets and finally decided to travel to earth and not any other planet. Where else can they find culturally and artistically endowed people but in Tamil Nadu and who else but the heroine Vani , who is a highly gifted village girl in the field of art, music and dance.

It is interesting to see how the science fiction drama has been finely tuned into a story which Tamil audience will adore. The exploits of the hero M.G.Ramachandran as he risks his life and rescues Vani after being kidnapped to the alien world, was a story line which definitely would keep the audience riveted to their seats.aani has reached the alien planet and is teaching the world to sing, or at least teaching the princess Rajini (Rajasree) dancing and singing.

Laser guns and a monitor similar to a TV set used by aliens to spy on activities on earth keeps the drama going. The hero meets his clone in that planet and is given anti gravity boots to be able to walk.salien planet. The film has all the basic ingredients of a sci fi film.

siu Biswarup Mukherjee a boy in his teens loved to dream about an extra terrestrial creature. His father scientist Avirup Mukherkee after his long and continuous effort successfully able to connect with them and Biswarup used called that creature "Friend." Ranju da and Sanju da a helpful neighbor of Avirup also cooperated and gave them a good support regarding there security and support. Some of the colleagues of Avirup tried a lot to spoil the experiment but strong determination and family support helped him to get the success. Later Avirup was kidnapped and few antisocial forced him to sell his product to some underworld people. The kingpin of the group was no one other that Avirup's boss Professor Brajamohan. In such situation, "Friend" became active and destroyed the evil and saved Avirup and his family. Unfortunately, all such incidents really made Avirup sad and he decided to send "Friend" back to his own territory.

Anti social elements after Avirupand his device, Biswarupisforced tosend the alienback fromearth.The filmmakes ascathing attack on the acquisitive, cruel, insensitive andviolent natureofMan.

One of the most popular and most expensive Indian film at that point of time was science fiction Tamil film Enthiran released in 2010 directed by S. Shankar starring Rajinikanth and Aishwarya Rai Bachchan with Danny Denzongpa, Santhanam and Karunas. Academy Award winner and Padamshri A.R. Rehman gave the music and Art Direction was done by Sabu Cyril. Among the highest grossing films of all times it won two National Film Awards and three Filmfare Awards.

Scientist K. Vaseeegaran (Rajnikanth) invents a super-powered robot, Chitti, in his own image ,however it --------does not get approval due to not having emotions and the ability to make rational judgment.

Overtime Chitti develops emotions and is virtually ready to be part of the world. The story takes a dramatic turn when Chitti falls in love with Dr. Vasi's fiancée Sana (Aishwarya Rai) and turns rogue. Chitti purposely fails in an examination to be recruited into the Army following which DrVasi breaks it into pieces and throws it into a dump. Chitti however rebuilds and is picked up by antagonist Dr.Bohra who introduces a red chip into Chitti which enhances its anger and ruthlessness. Chitti goes about with its other robots which he creates into a spree of death and destruction with police force unable to control it but eventually Dr Vasi controls it and removes the red chip which cools it.

In the courtroom Chitti finally shows its human touch by turning the case against Dr Vasi , in favour of Dr.Vasi by revealing a visual record of Dr Bohra introducing the chip in its body. Dr Vasi is released but Chitti has to be dismantled.

The film does highlight how an inanimate object like Chitti display superior human qualities like empathy and self sacrifice.Does this one act make Chitti human?

BENGALI LANGUAGE FILMS

In Bengali language the film 'Friend' is a science fiction film like ET. Scientist Avirup Mukherjee's son Biswaroop dreams about aliens and eventually his father is able to connect with one and Avirup calls him 'Friend'.

With anti social elements after Avirup and his device , Biswarup is forced to send the alien back from earth . The film makes a scathing attack on the acquisitive, cruel, insensitive and violent nature of Man. Written and directed by Satabdi Roy, the film has renowned actors like Tapas Roy, Soumitra Chatterjee and Satabdi Roy.

Another Bengali film "Patalghar" The Underground Chamber based ona sci fi story has a scientist who has built a musical device in his laboratory,, Patalghar" which can put people to sleep with its music. An alien from planet Nypcha who has come to earth and tries to steal the device is put to sleep by it. Hundred and fifty years later attempts are made by another scientist to discover the device with number of other people who also interested to acquire it.

After many adventures the device is finally found and everyone wants to possess it including the alien who has woken up from his sleep during this time.But no one gets the device which is destroyed. At the end the device is destroyed but the scientist and the alien travel to planet Nypcha. The film is directed by Abhijit Chaudhuri.

Professor Shanku O El Doradois a Bengali science fiction film of 2019 directed by Sandip Ray based on , "INakur Babu O El Dorado", a story by renowed director Satyajit Ray. Dhritiman Chatterjee portrayed the protagonist character of Professor Shanku. And featured Mithun Chakraborty, Jackie Shroff and Ashish Vidyarthi as lead characters.

This is a sci-fi fantasy adventure story where scientist and inventor Professor Shanku visits the heart of the Amazon forests in search of the mythical cityof El Dorado. Jackie Shroff stars as the antagonist who wants to take control of dangerous crocodiles for his own park.

Jole Jongole is a Bengali movie released on 2017and directed by Nitish Roy in which a scientist, wants to bring back prehistoric creatures, perished millions of years ago from the earth as also trying to enhance the intelligence level of the brain of crocodiles. He succeeds in creating a crocodile which has been genetically implanted with intelligence almost similar to that of dolphins. A greedy NRI wants to acquire the crocodile and build his own Jurassic Park and create a lucrative commercial venture.

The 2015 Bengali film 'Abby Sen' has time travel as its theme and the complications which ensue when a television producer travels from 2013 to 1980. It is interesting to see how does the television producer meet with the demands of a competitive professional career and personal life in both the situations.

The science fiction comedy film is directed by Atanu Ghosh and produced by Firdausul Hassan and Probal Halder. The actors in the film are Raima Sen, Chiranjeet Chakraborty, Bratya Basu, and Priyanka Sarkar.

Abby Sen is a 30-year-old television producer in the 2013. He is academically brilliant, has a strong background in science and watching science fiction films is his greatest passion.But his programmes on television fail to make a mark on the TRP ratings. Abby has lost several jobs and therefore keep the latest job loss a secret from his wife who tends to become hysterical. Abby meets a man who claims to be a scientist who has discovered a time-travel capsule. The scientist is ready to take him to a time when it was easier to get jobs and Abby travels 33 years back in time, to 1980. Managing professional and personal work is a tough job in itselfand that too in two different time period smakes it all the more interesting.

KANNADA LANGUAGE FILMS

In 2002 Kannada science fiction film "Hollywood" written bya ctor Upendra Rao and directed by Dinesh Babu was released claimed to be the first Indian robot film. It starred Upendra in a triple role as Surendra, Upendra and US47 (a robot) along with the Australian actress Felicity Mason as Manisha. The movie was shot entirely in Gold Coast, Australia. Upendra became the first Indian actor to portray the role of an android robot in a leadrole.

Surendra and Upendra in the movie are identical twins. Surendra is a research scientist in Los Angeles and Upendra is a filmmaker from India. Upendra comes to live with Surendra to make an Oscar winning Hollywood film. Surendra is in love with Manisha a blond neighbor.

However Manisha loves Upendra as he is smart and Surendra is shy.

Surendra's professor agrees to create a robot to fool Manisha into thinking it's Surendra with added charm and onc eManisha starts loving Surendra (robot) real Surendra will take over. However the robot develops human emotions (a love for Manisha) and with superhuman strength, it cannot be stopped when it tries to eliminate Surendra.

MALYALAM LANGUAGE FILMS

Malayalam-language science fiction film Karutha Rathrikal (DarkNights) released in 1967 directed by P. Subramaniam is an adaptation of Robert Louis Stevenson's\ DrJekyll and Mr Hyde. It was the first science fiction film in the history of Malayalam cinema starring Madhu, K.V. Shanthi,T.K. Balachandran and Rajasree.

Santhan, a medical practitioner, is in love with his cousin Vimala, with whom his marriage is fixed. Vimala's father, a banker had died under mysterious circumstances. Santhan develops a medicine, which transforms a person into a monstrous creature when consumed. He also invents the formula that reverts the person back to their original self. Santhan keeps this invention a secret. After reading his uncle's diary, Santhan realizes that other directors of the bank were responsible for his death. Santhan plans revenge on them by using his invention.

Vimala's cousin Mohan is in love with Vilasini, anight club dancer. The other bank directors influence Mohan, through Vilasini, and try to steal the bank documents from Vimala's house. At hat time a monstrous creature (Santhan) appears before Vilasini and threatens to kill her if she steals the documents.Thereafter the bank directors involved in the conspiracy are found murdered one after another. Vilasini discovers that Santhan is the one killing the bank directors. Mohan also learns about the secret that Santhan is the monstrous creature. Vimala's uncle Kochammavan comes to the city to conduct Vimala and Santhan's marriage. The police also arrive there in pursuit of the monstrous creature. Mohan reaches the spot and exposes Santhan, who then transforms himself into the monstrous creature in front of everyone. But before the police can arrest Santhan, he commits suicide.

Aathisayan is a 2007 Malayalam-language science fiction film directed by Vinayan, starring Master Devadas, Jackie Shroff, Mukesh and Kavya Madhavan in the lead roles. The film was inspired by the 2003 American superhero film Hulk. Athisayan marked the debut of Bollywood actorJackie Shroff in Mollywood.

Maya, a young television reporter, works for New India Television. Radharamanan, the Managing Director of the channel, has a fascination for Maya, who was brought up in an orphanage. Maya takes up the responsibility of bringing up a few orphans whom she has picked up from the streets. One of these kids is highly intelligent Devan. R. C. Shekhar, a scientist, lives very near their house, with his servant Damodaran. He is working on a project, which could help in making human beings invisible. On the very day that he successfully experiments his 'magic' potion on a rabbit, he is urgently called to the U.S as hisdaughter, who is studying there, meets with an accident. Meanwhile, Maya's friend and lover Roy,also a journalist, is in jail after having been framed in a murder case by some influential politicians including the State Revenue Minister Divakaran (RajanP. Dev), another minister Yunus Kunju, the Police Commissioner Shanmughan and the very influential Kuwaiti Nazar. With the help of the newly appointed Collector

Anita Williams, Maya shoots a video of these persons accepting millions of rupees as commission from some Arab businessmen.

Maya immediately goes to Radharamanan and shows him the clip. Radharamanan is thrilled and agrees to telecast it, after discussing it with the Board of Directors. Somehow Nazar, Divakaran and others come to know of this and are on lookout for Maya, in order to get from her the memory card containing the clipping. On the very same day that R.C.Shekhar goes to theU.S, Maya is abducted by Nazar and group.

Devan, who had seen R.C.Shekhar conducting the experiment on the rabbit, steals into Shekhar's lab and drinks the potion. He becomes invisible and sets out to save Maya. Devan does manage to save Maya, but gets shot and due to the scratch of the metallic bullet becomes a huge monster. He then goes and kills the villains who had harmed him and his family. He then leaves Maya and the children and makes his abode in the deep ocean, with an end note saying that "to fight for the Justice, Devan will come back as Athisayan".

The Malyalam film Android Kunjappan Version 5.25 is a 2019 sci fi film written and directed by Ratheesh Balakrishnan Poduval starring Suraj Venjaramoodu, Soubin Shahir, Kendy Zirdo and Saiju Kurup. The movie won three Kerala State Film Awards including Best Actor Award.

Bhaskara Poduval is a stubborn old man, residing in Payyanur with his son Subramanian (Chubban) a mechanical engineer. Due to his fear of dying alone he forces his son to quit jobs and continuously stay with him. However Subramanian (Soubin Shahir), getting a job in a Japanese firm in Russia he goes there. There he meets Hitomi, daughter of a Malayali father and a Japanese mother and married her. She tells him that to take care of his father during his last days she had taken help of a robotic nurse.

Chubban comes home with the latest Android robot version 5.25 to take care of his father .Initially hesitant, Bhaskara Poduval reluctantly starts accepting the Android robot and becomes attached to it. Poduval calls the robot Kunjappan and starts to see it as his own son making Chubban anxious and insecure.

He decides to quit his job and goes back home.One day Chubban's father leaves home with Kunjappan and goes to a forestnear his home. Chubban and Hitomi search for their father and Hitomi tries to kill Kunjappan and some people steal Kunjappan"s head . Kunjappan tries to choke to death

Chubban but his father Poduval saves Chubban and at the end endearingly calls Chubban"Kunjappan".

So the end of the movie clearly establishes that an android can never really supplant human relationships. It is the insecurities in humans regarding their relationships which makes them feel so.

Red Rainisa 2013 Malyalam science fiction thriller film, written and directedby Rahul Sadasivan. The film stars Narain, Mohan Sharma,Tini Tom, Shari, Devan, Leona Lishoy, Andrea Fortis and Sergio Kalei.Red Rain was produced by Sachin Sadasivan. A number of European artists and technicians collaborated in the film. A young researcher named Jay (Narain) who becomes convinced that the strange phenomena of red rain are a result of extra terrestrial life.The film was based on the Kerala red rain phenomenon. Jay searches for the mysterious events behind the strange deaths of cattles and bright lights in sky. He sets on a journey to find the truth and fears of something more frightening to come.

In the monsoon of 2001, a drenched Kerala had actually witnessed a strange phenomenon in its abundant showers –the colour red. Samples later concluded that locally proliferating algae

Caused the coloured rain. But among the many hypotheses debated at that time that the rains held proof of extra-terrestrial life.

Scientists from UK and India working on the Red Rain phenomenon including Chandra Wickramasinghe, Rajkuma rGangappa (Univ. of Glamorgan UK) Milton Wainwright (Univ. Sheffield UK), A. Santhosh Kumar (Cochin University India), Godfrey Louis (Cochin University India) had suggested after studying the microbes that the red cells found in the Red Rain could survive and grow after incubation periods of upto two hours at 121 degree Centigrade, experiments also showed that cells can replicate under high pressure at temperatures up to 300 degree centigrade. They said that flourescence behaviour of the red cells is similar to the extended red emission observed in the Red Rectangle planetary nebula and other galactic and extragalactic dust clouds, suggesting, though not proving an extra terrestrial origin.

Rahul Sadasivan met Professor Chandra Wickramasinghe from Cardiff University before making the movie.

TELEGU LANGUAGE FILMS

In Telugu language we have just one or two films which have a subject related to science fiction but with a strong fantasy element.

Aditya 369, released in 1991 is aTelugu film written and directed by Singeetam Srinivasa Rao which deals with Time travel. The film stars Nandamuri Balakrishna and Mohini along with Amrish Puri, Tinnu Anand, and Suthivelu. It received two Nandi Awards for Best Costume Design and Best Art Direction. It was dubbed into Hindi as Mission 369 and into Tamil as Apoorva Sakthi 369. While Aditya referred to sun, 369 was the serial number of the Time Machine.

The story talks of Prof. Ramdas a scientist who does experiments to invent the time machine at his home laboratory however is not successful. Meanwhile, Raja Verma, a high-profile thief steals antique pieces from the world's museums. His men steal a 16th-century diamond, belonging to the period of the Vijayanagara empire from the SalarJung Museum.

The robbery is witnessed by Kishore, a school kid who is trapped in the museumon his school excursion. He manages to escape from the robbers and gets rescued by Krishna Kumar. However, no one believes Kishore, and the diamond is replaced by its duplicate in the museum. Kishore gets to know that Prof. Ramdas is working on the time machine through his daughter Hema. One night, he along with other kids set off in that time machine to go to the day of the robbery.

Hema and her fiancé Krishna Kumar want to rescue them but they also get accidentally trapped in the machine. The time machine takes off and visits the past, to the court of Krishnadevaraya one of the greatest emperor of the Vijayanagara Empire who reigned from 1509–1529 and Krishna Kumar saves Simhanandini,a dancer in the royal court of Krishnadevaraya from raiders, and she introduces them to the Emperor.

Krishna Kumar surprises Krishnadevaraya by reciting the poem of his court's poet Tenali Ramakrishna before it's even written and explains to him that they have come from the future.

Though Krishnadevaraya finds it hard to believe, heo ffers them hospitality. Meanwhile a police constable who also happens to be trapped in the Time machine joins them. They see the stolen diamond in the possession of Krishnadevaraya.

Later, Simhanandini who loves Krishna Kumar accuses him of cheating her.However, Krishna Kumar is deemed innocent after a trial. To seek revenge, Simhanandini conspires with Senadhipathi to frame Krishna Kumar in the diamond's robbery.

Upon witnessing the diamond in Krishna Kumar's hand in a tussle with Senadhipathi, Krishnadevaraya sentences him to death. Though on the day of the execution, Krishnadevaraya who believes Krishna Kumar's innocence saves him. This is later confirmed to him by Ramakrishna who witnesses the diamond being robbed by Senadhipathi. Krishna Kumar, Hema, and the constable escape and board the time machine which sets off again. The machine takes them to the year 2504, adystopian world destroyed by radiation after the end of the Third World War. Scientists of that era already know about their arrival and warmly receive them. During their stay, they also watch the news from the year 1991. It reports that the diamond is retrieved from Raja Verma with the efforts of Krishna Kumar but he is killed in the process. The malfunctioning time machine is now repaired and when the environment begins to negatively affect them, they leave. The time machine brings them back to the present.

Raja Verma, who kidnaps Prof. Ramdas and Kishore to get hold of the time machine locates it abandoned on a hilltop. Krishna Kumar rescues them and combats with Raja Verma in the time machine. It is destroyed in the feud and both are reported dead. But Krishna Kumar who jumps off the cliff moments before it explodes is saved and joins his family. Renowned singers S. P. Balasubrahmanyam and S.Janaki's songs and legendaryIlaiyaraaja's music greatly enhanced the appeal of the movie.

The first space based science fiction film of Telugu is supposed to be Antariksham 9000 KMPH which was released in the year 2018 directed by Sankalp Reddy. The film stars Varun Tej, Aditi Rao Hydari and Lavanya Tripathi.

Basically the film shows crisis created by malfunctioning satellite and the hero a scientist helps in restoration of the satellite with a love story also woven in.

Dev (Varun Tej) who is an astronaut who is living incognito for past five years as a teacher following an accident is roped in to fix a moon orbiting satellite with which he had been involved as it threatens to disrupt the communication system worldwide.

Four astronauts Dev, Riya, Karan (Satyadev) and Sanjay (Raja Chembolu) are launched into space and are eventually able rectify the problem and save earth from communication disaster.

Actors like Varun Tej, Lavanya Tripathi, Aditi Rao Hydari, Srinivas Avasarala, Satyadev and Raja Chemboluput life into the characters and the film , however as many other Indian sci fi flicks the serious scientific storyline is perfunctory compared to the heavy dose of romance , patriotism and family drama as any typical Indian movie with high quality music and sets.

MARATHI LANGUAGE FILMS

Marathi film Phuntroo is a science fiction film starring Madan Deodharand Ketaki Mategaonkar which seems to have been inspired by Hollywood film 'Her' where the protagonist gets emotionally attached with an OS (operating system).

Directed by Sujay Dahake 'Phuntroo 'has Vira (Madan Deodhar) as the protagonist who is an introvert and does not have any friends. He loves his senior Anaya (Mategaonkar) but he doesn't have the courage to express his feelings to her who is close to Navneet (Waichal). Even as Anaya and Navneet come closer, Vira expresses his love to Anaya but is rejected.

Annoyed at the refusal, Vira' create's a hologram of Anaya demonstrating his technological skills and then wants to give her a physical form. Navneet finds out his secret.

Having a virtual copy of Anaya does bring out all the moral and philosophical issues including the relationship between Artificial Intelligence and Humans.Canacopy of Anayabe Anaya and can we have a similar emotional relationship with AIas with Anaya?Can a machine"love"?

GUJARATI LANGUAGE FILM

In Gujarati language the first ever sci fi film is a successful äction thriller with lot of comedy and drama called, "Short Circuit"

The hero is a computer maintenance guy Samay (DhvanitThaker), who loves Seema (Kinjal Rajpriya) a news anchor.

Evil scientist Dr. Parthsarthi (Utkarsh Mazumdar)who is involved in creating a wormhole in his laboratory threatens to kill Seema if there is any attempt to stop him.

Seema is shot and killed by a masked rider as soon as she steps out her office.

Samay who was planning to propose to Seema get drunk with his friend Bhopa (Smit Pandya)

Samaywas planning to propose Seema..Samay gets drunk with his friend, Bhopa (Smit Pandya) get drunk at night and strangely both of them see Aurora Borealis in the sky in India.

When Samay wakes up the next day, he starts to live the same day again and again. How he able to save Seema and how he gets out the time loop forms a very engrossing movie.

Produced by Twilight Productions the film is directed by Faisal Hashmi and written by Faisal Hashmi, Bhargav Purohit and Mohsin Chavada.

ODIA FILMS

The 2009 film " Keun Dunia Ru AsilaBandhu" is claimed to be the first science fiction film in Odia language dealing with the relationship between a young Alien (Alpha 10) and a physically challenged young boy.

The film is made by Odia film director Sanjay Nayak with renowned actors like Sabyasachi, ArchitaSahu, besides Mihir Das, Budhaditya Mohanty, Minaketan, Biren Mishra, Babli, Priyanka Patnayak among others.

The story line is that a business man cheats his close friend an astrologer and takes away his property. The business man's wife gives birth to a daughter but in the hospital itself the businessman pays money and gives the daughter away to a snake charmer and takes his newborn son.

The astrologer has two sons and his younger son (Sabyasachi) is physically challenged who is bullied by his brother and other boys. He does not got school and loiters around and his helpful and even helps in cleaning the area and surroundings.

Sabyasachi meets the snake charmer's daughter and becomes good friends. One day both of them in a forested area close to her residence find an alien lying on ground who is wearing a watch that is emitting a ray. The alien shone the ray in the physically challenged boy's arm , as soon as the ray it touches his arms it starts functioning normally.

He takes the Alien to his house and there also as soon as the ray from the watch touches his mother who is paralysed and moves around in a wheel chair , she become normal. They keep the Alien in their house.

In the meanwhile news is flashed all around that an alien ship had landed on the earth in their city and one of the Alien is stranded unable to go back. A search in on by the authorities.

The Sabyasachi's elder brother is in police and is tasked to lookout for the Alien being completely unaware that Alien is staying in his house being hidden in the almirah by his younger brother.

By word of mouth it spreads that Alien is in the astrologer's house, the elder brother then searches for the Alien in his own house also but is unable to find him.

Alien has been hiding in a broken almirah and is found to be in a breathless state ,the Alien tells Sabyasachi that his energy levels are coming down and unless he gets food (chemicals to be found in Space Research Centre) he will die.

Sabyasachi wearing his brother's police uniform gains entry into the Space Research Centre and stealthily takes away the required chemicals.

The elder brother finds the Alien but on request of his father making the plea that he had cured his mother and brother, does not inform authorities.

It is decided to take the Alien to a remote area.

After a few days Alien again needs food and again chemicals are brought from the Space Research Centre.

The businessman with number of anti social elements comes to take the Alien forcibly away and there is fight among both groups with the Alien being involved in it too. The Alien uses his ray to defeat the businessman and his toughs.

The film ends with a number of space ships arriving at the desolate spot and one of the spaceships opens from which alight two Aliens who come and take the little Alien away in their spaceship.

CHAPTER 8
SPECIAL EFFECTS BEFORE COMPUTER AGE

One of the major difference between normal films and science fiction films is that it is replete with special effects which are required with its scripts usually dealing with spaceships and aliens, high tech gadgets, locales which may be from any part of the universe, time travel and futuristic scenarios, astonishing action sequences, giant sized characters to tiny miniatures.

Audiences pack to see science fiction films primarily to see all the special effects which is virtually like a magic show on the silver screen. From the beginning of science fiction films special effects have been the sine qua non of SF films.

One of the earliest special effect to be seen in a film though not a science fiction film was in "The Execution of Mary Stuart " in 1895 when using the trick of a simple stop motion an execution was depicted.

The particular scene was filmed till the actor doing the role of the queen (who was to be beheaded) came to the spot to be executed .All other actors were asked to stand still and the camera was stopped , the queen was replaced with a mannequin and then as the camera rolled again the executioner chopped the head of the mannequin. This little camera trick of "stop motion" made it so realistic that some people during those days are supposed to have thought that a person had actually given one's life. The 18-second-long film was produced by Thomas Edison and directed by Alfred Clark. At this time the special effects were produced by just a simple technique of editing or cutting and assembling of shots by joining the films. To bring about transition from one shot to another you could have fades, swipes, dissolves or simply cut from one shot to another if required. Techniques like this dominate even today. In the first science fiction film „ "Le Voyage Dans La Lune " (A Trip to the Moon) in 1902 directed by George Melies,a renowned French magician known to be the father of special effects, a number of special effects were used including stop motion, double exposure, split screens. Double exposure gave the photographers enormous possibilities to create special effect.

In early days of photography when they had to expose the film for 20 or 40 minutes the person being photographed had to be still for the entire time, if the person shifted from his position, you would get a double exposure. The final photographs would show the same person in two different positions. Georges Melies has written about this accidental discovery himself ,"The camera I was using in the beginning, a rudimentary affair in which the film would tear or would often refuse to move, produced an unexpected effect one day when I was photographing very prosaically the Place de l'Opera . It took a minute to release the film and get the camera going again. During this minute the people, buses vehicles had of course moved. Projecting the film, having joined the break, I suddenly saw a Madeleine-Bastille omnibus change into a hearse and men into women. The trick of substitution, called the trick of stop-action was discovered..." Even though initially it happened accidentally later it was used a great deal by film makers to have a duplicate of the same person sitting opposite himself, or the same person in several places by multiple exposure in different positions. Once double exposure and multiple exposure was discovered cinematographers continued to use the technique in variety of ways to show more than one person, scene or object on the single shot depending on their creativity by double or multiple exposure of the same film. Another often used technique was the „"split screen" technique which Melies introduced in which he covered half the film and then shot his image and then rewound the film covered the other half and took another shot When the film was developed the two images appeared side by side in real time. It allowed same actor to perform opposite himself.

Another technique known as matting used was to mask certain sections of the film so that two images can be combined into a single shot. In 1898 film „ "Four Heads are Better Than One" by Georges Mélies, Melies blacked out or matted out parts of the frame using black paint and a piece of glass. He would black out one part so no light would reach the film and then rewind the film and matte out everything else.

The resulting multiple exposure combined the shots into one frame. So four heads in one shot was actually a cleverly done four separate exposures on the same film.

These special effect techniques also allowed a person to disappear or a person or object to suddenly appear from nowhere, or even an object to suddenly grow in size simply by taking shots closer to the object or far away and by clever editing make the object suddenly seem large or small.

Another important trick used a great deal by cinematographers almost routinely was "forced perspective". Actually a large number of sci fi films use miniature models of sets showing the alien cityscape with futuristic and out of the world architecture, or a house or a furniture or a natural scene as the requirement may be and by using „ "forced-perspective" give the audience the feeling they are seeing the real thing.

The forced perspective technique uses the natural optical illusion of the eye to make an object appear farther away, closer, larger or smaller than it actually is primarily by use of camera and the background props.

As our eyes perceive by looking at the correlation and perspective between the size and scale of objects and the distance at which they are from us, cinematographers can fool us completely placing the camera nearer or further from the object and at particular angles to give a different picture of reality. A hanging man's legs with miniature cars at the bottom can be made to look as if he is hanging from the top of the building.

At times moving pieces of sets on mounted movable platforms can give us the impression of actual movement taking place outside. Like parts of building falling down in an earthquake can be shown realistically by using models and by filming them in the right perspective. Huge storms in the ocean are sometimes just created in a water body in the studio premises using fan blades.

Puppets and clay models are a favourite of special effects artists.

In the Lost Word film of 1925 giant monsters allosaurus and brontosaurus were created on the scene by using small puppets. By filming frame by frame the movement of the limbs and gaping mouth of the animals , the giant dinosaurs could be shown to be moving menacingly towards the actors who could be shot separately and then by editing inserted into the film showing them to be scarily close to the dinosaurs or desperately running away from the animals to produce high drama for the audience.

Kong, the giant gorilla, was actually not so "giant." He was a mere 18 inches, a model, covered with rabbit hair. The scene with actress Fay Wray at the top of the Empire State Building was filmed one frame at a time using stopmotion photography by visual effects artist Willis O'Brien and his crew. The producers filmed Kong and Fay Wray, separately. They then projected the two films together to create the effect of Fay Wray in the grip of Kong.

The Optical printer was an important development in film making as it not only allowed making duplicate of the film but served also as a special effects tool. The optical printer has a projector at one end and a camera at the other.

By running film through the projector, and rephotographing it with the camera, a duplicate or several duplicates of the original film can be created .

By zooming the camera or the lenses you can change the size and orientation of the final image. You can also slow and hasten the speed of the film by reducing or increasing space between shots , two or more images can be placed in single frame or other required images shot separately can be inserted where required in the final film.

Earlier to make the actors act against a specific background required blacking out all but the actors on the top film, then blocking out where the actors would appear on another film where only the background would remain and then printing them both onto a third roll of film.

The first film to use a blue screen behind the actors (which made it easier to print only the actors on the film) was The Thief of Bagdad (1940). Using this method, the film would be developed with a number of color filters to ensure that the blue background would disappear, while the actors would show up and any other required background could be inserted.

Movies including well known ones (BenHur, Star Wars, Titanic) have always used backgrounds made of giant glass panels known as matte painting to show a landscape,cityscape, or any architecture or natural scenery, or crowds.

Many matte paintings give the required background while the actors act in front of it. The first time this was done was in the 1907 film Missions of California, which used a massive matte painting of crumbling Missions. Alfred Hitchcock used glass pane matte paintings throughout the 40s and well into the 60s. They were prominently used in North by Northwest and The Birds.

But where a moving background is required, live footage is projected on a screen behind the actors then filming both the actors and the background film at the same time, in the same frame, which keeps the actors moving as well as the background action taking place simultaneously.

In filmmaking for a long time when the audience saw the car being driven actually the car and its occupants including the driver are stationery with the driver just faking movements on the steering wheel, the buildings race past on a large projected screen behind making the audience feel that the car was actually moving on the road. Mixing and matching with some real shots on the streets would make it very difficult to find out which is fake and which is true while a car chase takes place.

Actors can be placed under a building or any other artifact by placing a mirror on top part of the camera frame reflecting the model of the building or artifact and then shooting the actors against a wall. The eventual film will have the actors performing under the building or artefact.

As early as 1927, the film Metropolis directed by Fritz Lang used a technique known as the Schüfftan process which became very popular and was used as a special effect method for years afterwards.

This technique was especially useful if one had to show a group of actors inside a specific building. For using the Schufftan process a plate of glass is placed at a 45- degree angle between the camera and the miniature model of buildings. The outline of the area into which the actors would be performing is traced out and transferred onto a mirror. Besides this bit of reflective surface, all other reflective surface of the mirror is completely removed leaving only transparent glass. Actors would be placed strategically where they would be reflected on the mirror while the building or set would be seen through the plain glass.

The movie camera would then take the shots simultaneously of both the actors reflected on the mirror alongwith with model of the building or set through the transparent glass. Then the final film would show the actors inside the building.

Schüfftan process was used by many film-makers, including Alfred Hitchcock, in his films Blackmail (1929) and The 39 Steps (1935), and as recently as The Lord of the Rings: The Return of the King (2003) .

In 1918, a new process was patented by Frank Williams. The Williams Process placed actors in front of black backgrounds and filmed them.

The actual required backdrop was separately shot on another film. Using both films and an elaborate process of making negatives and positives, one could eventually get a single film to show the actors moving before the required backdrop. The Williams Process was used throughout the late 20s and into the 1930s.

The process was used in The Invisible Man. Actor Claude Rains wore a black velvet suit underneath his costume. The Williams Process was the predecessor to blue and green screen compositing done today.

One of the main technology used in special effects is compositing or "chroma keying," which means that visual elements from separate origins are eventually shown on the same shot. This visual effect technique requires filming with a blue screen as a background and then the entire background is removed and replaced with the required inputs which may have been shot at some other place.

For example you can shoot two knights clashing with swords taken in front of the green screen and later the screen is replaced with background of matte painting on glass of a jungle or set of a palace as you like.

Science fiction films primarily use visual effects because it allows them to create imaginary universes, habitable planets, alien architecture, besides all kinds of gadgets and actions that would be impossible to show in the real world.

In film making, visual effects create images, environments, objects, creatures, that do not physically exist in real life which are largely required particularly for science fiction films like innovative gadgets, novel environments and settings, extraordinary happenings like people or objects disappearing or appearing out of the blue and all kinds of creatures of all possible shapes and sizes.

Without use of computers one of the most memorable scene in science fiction film history takes place in Alien (1979) when the alien bursts out from John Hurt's chest. The scene is supposed to have shocked a number of viewers in cinema halls. Interestingly it is reported that the co stars were also completely taken aback and gave real life reaction when the shot was taken because they were not also made privy to what is going to come out of the belly. The actress who was suddenly sprayed with blood actually fainted during the shooting.

The prosthetic body filled with organs got from butcher shop with compressed blood machine containing six gallons of red liquid was set off by small explosives by the hidden crew. As the chest burst, blood splattered over and the grotesque looking creature shot out and sped away (actually a puppeteer held it with a stick and moved it and a air hose helped its tail to move about).

Similarly, in 2001 : A Space Odyssey 's famous star gate sequence the psychedelic light show at the end was done without use of computer. The special effects were created by making the camera move down a track in a darkroom towards a sheet of glass painted black with a narrow four foot slit down its middle. Behind the glass were a wall of coloured gels which also moved left to right. As the camera would close in towards the slit after every 60 second exposure the wall of gel would be moved slightly.

The final effects were as if one was just moving continuously through fast changing coloured lights of different shapes. Through the sequence you had the camera automatically focusing from 15 feet to one inch and the lens was kept open all through providing a "wow" special effect. Doug Trumbull who did the special effects said "We wanted the audience to feel like they were actually going to space and on this adventure themselves, not just through the characters, but participating." According to Trumbell himself in the Stargate sequence "we transcend time and visually seem to be in another dimension". Trumbel's exposure to time lapse photography and streak photography on the lighted roads at night seemed to have helped him in creating the stargate sequence.

In 2001 : A Space Odyssey there a scene where an astronaut is jogging sideways around a centrifuge in space. This stunning effect was achieved by building a gigantic hamster wheel which rotated around a single camera. Another remarkable visual effect which would remain in the viewers mind will be the actors walking in zero gravity in the same film. For this all the fixed fixtures like the chairs,desks,panels all were bolted to the wheel while the entire spacecraft was seen to be rotating at a speed of three miles per hour. One camera was mounted stationary to the set, so that when the set rotated, the camera went right along with it. The screen would then be able to show the actor going up the wall around the top and down the other side . In the second type of shot the camera, mounted on a miniature track, stayed with the actor at the bottom while the whole set moved past him.

In Jurassic Park the T-Rex, was not a computer graphics image but was a 9000 pound animatronic hydraulic powered machine specially built for the film.

The science fiction film Fantastic Voyage got Academy Award for Best Achievement in Visual Effects for being able to show miniature human being travelling into the bloodstream of a human body for which a full-size hightech navy submarine actually was used. The interior of the body was created by using large, highly-detailed sets of various body parts (i.e., the brain, the heart). The brain was 100 foot long and 35 foot high, thealmost realistically pumping heartmade of rubber and latex was 130 feet wide and 30 feet high, while the lung sac was constructed so that it would seem to breathe contained 300,000 cubic feet of surface area made or resin and fiberglass,thepulmonary artery was8 feet in diameter and 40 feet long.

The military complex which was created where the submarine Proteus was shrunk and the laboratory was 100 feet by 300 feet costing nearly $ 1.25 million to build. Proteus was 23 feet by 42 feet, weighed 8,000 pounds and cost $ 1,00,000.

An optical printer was used to shoot matte paintings of smaller and smaller size of submarine Proteus against the laboratory background to create the impression of shrinking of the huge submarine.Blobs of red blood cells were nothing but coloured Vaseline and oil.

The aliens landing in Close Encounters of the Third Kind was a tense and spectacular moment in the movie. It was created by using a four feet high and five feet wide fiber-glass model . The other worldly look was brought about by just using neon tubes and bulbs lighted by fiber optics flickering in tandem with the music. The UFO's glow and music did give the real feel.

In the painstaking way in which the special effects are prepared and created provide a realism which is not possible even in CGI which is being routinely used in films of today. The feel of authenticity in film with non CGI special effects continues to remain and directors from time to time like to go in for it which may be a more expensive and time consuming proposition like handloom made to order shirt compared to a machine made shirt.

CHAPTER 9
THE TECHNIQUE OF COMPUTER GENERATED IMAGERY

Computer-Generated Imagery (CGI) is the creation of still or animated visual content with computer software.

The history of CGI goes back to the 1950's, when mechanical computers were repurposed to create patterns on to animation cels (transparent sheets on which drawing is made) which were then incorporated into a feature film.

The first film which used CGI was Alfred Hitchcocks Vertigo (1958).

John Whitney programmed these graphics using a computer. A pendulum (which contained pressurized paint) was placed above a drawing surface that was attached to a platform. The platform was moved by the computer according to mathematical equations as the pendulum swung back and forth across it. This created precise spiral designs.

In the opening sequence you can see these spiral designs change shape. These changes were created by altering the formulas for each frame that was drawn with the computer/pendulum combination.

It was in 1972 a short film was made showing a 3-dimensional moving hand created by a computer .The film,, "A Computer Animated Hand" "was created by Edwin Catmull and Fred Parke. Edwindrew 350 triangles and polygons in ink on his hand and then digitized it (converting information into computer readable format) and getting the output in line drawings and laboriously animating the data in a 3D animation program that he wrote on a computer, to finally create the film.

But it was clearly one of the early examples of computer graphic. Three dimensional computer graphics are today used to create characters, scenes and special effects in films, television and games. The technology is also used in everything from advertising, architecture, engineering, medicine,virtual reality.

CGI is used extensively these days in films because it turns out to be much cheaper than building huge sets or miniature models or hiring large number of extras for crowd scenes. Secondly CGI helps avoid the risks involved in filming dangerous sequences usually done through duplicate actors and stuntmen.

Three-dimensional graphics software over the years have improved a great deal and can create everything from simple primitive shapes to complex forms made from flat triangles and quadrangles.

Today filming is also done digitally and therefore computer generated graphics can be introduced into the digital film footage by using a technique called composting.

Compositing is a process in which visual elements from separate sources are combined into a single image, creating the illusion that all the different elements are a part of the same scene.

So, after creating through computer generated imagery a scene of a building blowing up due to bomb blast leaving huge clouds of smoke and fire, as a backdrop , one can place shots of hero and heroine running away from the site of the blast by composting in front of that background.

This is done using chroma key techniques and green screens. What this means that live action shots of th hero and heroine running away is taken in front of a green screen. This us done post production using video editing software. You can remove the green screen background and it will become fully transparent. The process is known as keying or chroma keying.

The backdrop is entirely replaced by alternate background video or CGI. In this case final edited version will show hero and heroine running away from the building on fire due to bomb blast.

Computer-generated imagery (CGI) encompasses both static scenes and dynamic images, while computer animation only refers to the moving images. The hero/heroine have to avoid wearing clothing that matches the backdrop or they will also disappear. Interestingly one can show only a floating head or a head coming out of a box by making an actor dress entirely in green. In post production chorma key will result in the actor becoming transparent leaving only the head.

It was way back in 1961, that a 49-second vector animation of a car traveling up a highway at 110 km/h was created at the Swedish Royal Institute of Technology on the BESK computer.The short animation was broadcast on November 9, 1961 on national television.

In early years of CGI a ten-minute computer animated film by Charles Csuri and James Shaffer was awarded a prize at the 4th annual International Experimental Film Competition in Brussels, Belgium and is in the collection of The Museum of Modern Art, New York City. The subject was a line drawing of a hummingbird for which a sequence of movements appropriate to the bird were programmed. Over 30,000 images comprising some 25 motion sequences were generated by the computer.

Today there are computer software which will generate for a movie scene required number of public to fill a stadium or large number of soldiers for a field of battle.There are also computer software to make natural background of grass and trees and flowers, hail and snow depending on your requirements which are almost like real images.

Live actors today can be naturally integrated with the computer generated people.You can have live horses moving along with computer generated horses. The obvious benefit is the CGI animated character can do tasks which would be very risky or difficult for the human actor or live animal to accomplish. Both can be so well integrated that it maybe difficult to distinguish which is CGI and which is not.

The CGI looks real because a lot of details are worked out, even for the large crowd scenes with each person 's complete body geometry, hair and beard and the way they wear their clothes and where they have the folds, what type of clothing they have with all possible variations taken into account can be created and distinguished separately.

For a winter scene the size, shape of the ice flakes including the light and shade on the ice and the required level of shine can be worked out precisely.

In a sense the graphic artists doing background are actually painting in the detail of the background as the matte painters or animation artists used to do earlier only in CGI it is virtually done and there is a lot of computer technology involved to produce the realistic image. However it rules out location and outdoorshooting to a great extent as lot of location can be inserted as part of post production work. The details which can be inserted due to power of the computer are of a much higher order than would have been otherwise possible besides taking much lesser time.

Not only can CGI create large crowd, the movement of the limbs, the face, the mouth, even the fingers of different members of the crowd all seeming genuine and real and different from one another.

With developments in CGI, Virtual Reality and Artificial Intelligence, Gaming, a new and vast field of virtual cinematography and virtual production is nowopening up with the possibilities of digital actors and AI directors.

To get the movement correctly on CGI what is usually done is an actor s movement is digitally recorded called "mocap" motion capture, like getting hundreds of variations of the movement or action required for the scene which is then transferred to a computer-generated 3D model. Once you have movements of the actor on the computers digital 3 D model then this virtual actor can be made to perform any awe inspiring action which will look absolutely real and matching with that of the actor.

The same process can also be done to get little details of facial expression of the actor by recording an actor's facial expression, called "performance capture." For motion capture an actor makes the motions wearing a motion-capture suit covered in special markers that a camera can track and record.

In case of performance capture, dots are painted on the actor's face which are recorded while making the expression. The data captured by the cameras is then mapped onto a 3D skeleton model using motion capture software.

All the individual soldiers in the battle scene thus act a individual and this is precisely what makes them seem real. It was Disney released "Tron"(1982) directed by Steve Lisberger a sci-fi adventure about an alternate electronic reality inside a computer which for the first time used 3 D, CGI and early facial animation.

Tron" received an enormous amount of advance publicity for heralding a new era in film making, in which computer images would replace models, miniatures, puppets, mattes and even sets and stunt men.

Terminator 2 Judgement Day, in 1991 was one of the the first realistic human movements shown on a CGI character. Terminator 2 also used high degree of sophisticated CGI technology to create the liquidifying and solidifying metal cyborg who after being shot leaving of hundreds of shattered fragments melting and flowing together and then coalescing and coming together.

Rendevouz in Montreal" was the first 3D film generated film involving virtual actors Marilyn Monroe and J Humphrey Bogart.

Death Becomes Her in 1992 was the first film to use the CGI software to depict human skin.

The twisting Meryl Streep's head around its neck and punch a huge hole in the abdomen of Goldie Hawn so that one could see through the hole. Both these special effects made a great impact on the audience.

Avatar"2009 was the first full-length movie made using performance capture to create photo-realistic 3D characters and to feature a fully CG 3D photo-realistic world'.

Few films have had as much influence on movie effects as The Matrix released in 1999 which got Academy Award for Visual Effects. Its most iconic scene is a frozen moment that has become known as 'bullettime', in which Neo (Keanu Reeves) dodges bullets fired at him by an agent, while the camera circles around. The sequence still captivates the audience.

Special effect tknown as bullet time is a visual effect whereby the passage of ime is slowed down so that any observer can see individual bullets flying throughout the scene . There is a still or action stops or virtually slows down as the motion continues, the camera rotates around the still object.

The effect is achieved by a set of still cameras surrounding the subject which are activated

simultaneously. The pictures in the still cameras are then displayed consecutively and spliced into movie frames, creating the effectof a single camera moving around a scene either frozen in time or moving in extremely slow motion.

In this particular case the bullet time effect here is slightly more complicated because instead of firing the cameras simultaneously, the visual effect team fired the cameras fractions of a second after each other, so that each camera could capture the action as it progressed, creating a super slow-motion effect. When the frames were put together, the resulting slow-motion effects reached a frame frequency of 12,000 per second, as opposed to the normal 24 frames per second of film.

In a film like Interstellar directed by Christopher Nolan where they had to show blackhole, computer graphics helped visual effect artists to generate realistic images of the hole and its gravitational lens by setting three key parameters: rate of spin, mass and diameter. Interestingly theoretical physicist Kip Thorne was also on board to provide the maths.

It was a proud moment for science fiction films when scientists released the first real black hole photo located 55 million light years away from Earth in the Messier 87 galaxy and found it to have striking similarity to the one shown in film.

Christopher Nolan has said in an interview "We hoped that by dramatising science and making it entertaining for kids we might inspire some of the astronauts of tomorrow."

The basis of CGI superb bright and sharp coloured imagery lies in the ability of the computer to assign values for brightness and color (a proportion of red, blue and green) to each point or pixel on a video monitor. The vividity and richness of the environment is worth seeing particularly in CGI based films like Avatar.

The transformations of the T-1000 involved a program called "morphing" (from metamorphosing) by which one shape is transformed into another slowly as whenT-1000 changed into Sarah Connor.

One of the highest grossing films Avatar by James Cameron creates a strikingly beautiful world using the CGI technology.

For the avatars, full-size models of generic Na"vi male and female characters were built and laser-scanned. For the rich and detailed vegetation full of exotic flora and fauna, 3000 separate plants and trees, were created for depicting Pandora jungle in a realistic fashion. Every tree, leaf, or even rock was individually created with the required light and shade.

With developments like Digital Actors, Virtual Reality, Virtual Cinematography, Virtual Production, Artificial Intelligence, Video Games,the CGI of the future may not be anything like it is today.

CHAPTER 10
PREDICTIONS BY SCIENCE FICTION WRITERS

Uncannily science fiction writers have been able to predict several future developments in human society much before its time while writing their imaginative works.

It may seem unbelievable but while taking their creative flight or immersed in deep thought they are able to foresee the future many many years ahead. It almost seems that they are able to tap into the some kind of universal Akashic record which keeps encoded all happenings past present and future.

It is not that all science fiction writers are able to do that, but some have had the knack to be able to predict the future almost precisely.

While astronomer Johannes Kepler had envisaged lunar travel in his Somnium (The Dream) written in 1608, the idea was so strange at the time that Kepler chose to have demons transport his protagonist.

In 1638, Bishop Francis Godwin had a similar flight of fancy: his protagonist in The Man in the Moone hitched a ride with migratory birds.

But in The "Other World : The States and Empires of the Moon"" , a 1657 early science-fiction story by French author Cyrano de Bergerac, the protagonist makes a machine that launches when soldiers fasten fireworks underneath it:

"I ran to the Soldier that was giving Fire to it… and in great rage threw my self into my Machine, that I might undo the Fire-Works that they had stuck about it; but I came too late, for hardly were both my Feet within, then whip, away went I up in a Cloud."

This was perhaps the first rocket using scientific principles catapulting an object into the space, the Sputnik sent by the Russians to space would come 300 years later in 1957.

Even though Jules Verne gets the credit for imagining a full fledged working submarine under the ocean, following the huge popularity of his novel 20,000 Leagues Under the Sea (1870),not many know that more than 200 years before the publication of Jules Verne's novel, a submarine was also mentioned in The Blazing-World (1666), a book about a satirical utopian kingdom, written by Margaret Cavendish, the Duchess of Newcastle.

Cavendish's protagonist talks to sentient animals about various scientific theories, including atomic theory, before travelling home in a submarine when she hears that her homeland is under threat. The Blazing World also provides a very early reference to submarine technology, and the possibility of underwater exploration and warfare.

Machine-automated language is referred to by Jonathan Swift in 1726. He critiqued the so-called scientific literature of his time, which was not always the result of rational thinking. Consequently, when Swift described an —engine‖ that could form sentences, he was satirizing the arbitrary methods of some of his scientific contemporaries: the most ignorant person, at a reasonable charge, and with a little bodily labour, might write books in philosophy, poetry, politics, laws, mathematics, and theology, without the least assistance from genius or study‖.

What Swift may not have realized was that his ensuing description of a machine containing all the words of the language spoken in Lagado, a fictional city, is one of the earliest known references to a device broadly representing a computer. Nowadays, computers are able to generate permutations of word sets, as Swift envisaged.

H.G.Wells who is considered to be father of science fiction conjured many futuristic visions in his novels some haven't (yet) come true like a machine that travels back in time, a man who turns invisible, and a Martian invasion that destroys southern England. But then some of his predictions not having come true so far do not in any way take away from the fact that in his novels he seems to have some kind of a precognition of the world of the future.

In his book Men Like Gods (1923), he builds a futuristic utopia years ahead of his time where people communicate exclusively with wireless systems that employ a kind of co-mingling of voicemail and email-like properties.

In Utopia, except by previous arrangement, people do not talk together on the telephone, he writes. —A message is sent to the station of the district in which the recipient is known to be, and there it waits until he chooses to tap his accumulated messages. And any that one he wishes to repeat can be repeated. Then he talks back to the senders and dispatches any other messages he wishes (exactly like today's answering machine on the phone or email). The transmission is wireless.

In "When the Sleeper Wakes" (1899), the protagonist Graham takes a drug to cure insomnia and gets up in the year 2100, after two centuries of slumber to a dystopian London. He finds many novelties like the audio book, airplane, and television— it being a dystopian society there is oppression and social injustice. The London that Graham knew was not there Anymore.

The London of the future, with a population of 33 million, is a vast, claustrophic metropolis with countless levels, connected by walkways. It has wind-wheels on the roof; huge flying stages for the aircraft of the future; while kinetelephotographs allow words and pictures to be projected around the world.

The railways are not there anymore but what you have is hundred yards —wide roads, made of toughened glass called Eadhamite, along which vehicles on rubber wheels sweep along at high speed – like the automobiles of today.

In the The Island of Dr. Moreau (1896) by Dr Wells the mad scientists Dr Moreau conducts human-animal hybrid experiments. He creates the Leopard-Man and Fox-Bear by surgical transplants and blood transfusions.

Scientists today are working towards the day when animal organs could serve as long-term transplants for human patients.

Naturally, Wells has portrayed Dr.Moreau's experiments as a warning for Man not to play God, which warning holds true even today as scientists are wary o go ahead through many highly controversial genetic experiments.

In the _ "War of the Worlds" written in 1898 by H.G.Wells , he shows Martians causing devastation by unleashing Heat-Ray, a super weapon capable of incinerating helpless humans with a noiseless flash of light, about 60 years before the laser was invented. Dr H.G.Wells wrote, "Many think that in some way [the Martians] are able to generate an intense heat in a chamber of practically absolute non-conductivity. This intense heat they project in a parallel beam against any object they choose, by means of a polished parabolic mirror of unknown composition, much as the parabolic mirror of a lighthouse projects a beam of light."

Wells also could foresee that splitting of an atom would cause global destruction and then lead to peace and nations coming together almost like what happened in real life, dropping of the atom bombs ended World War II but also led to creation of United Nations. Wells too wrote about a unified world government being created by the survivors of the war in which atom bombs are used in ''The World Set Free'' (1913).

In Wells's novel bombs kept on being triggered for months . He also warned about possible nuclear proliferation and the dangers of these weapons getting into the hands unauthorized persons.

'Destruction was becoming so facile that any little body of malcontents could use it; it was revolutionizing the problems of police and internal rule.

Before the last war began it was a matter of common knowledge that a man could carry about in a handbag an amount of latent energy sufficient to wreck half a city,' he wrote.

H.G. Wells predicted the creation of war tanks in his short story "The Land Ironclads".

Wells predicted that a new type of bomb fuelled by nuclear reactions would be detonated in 1956. It happened even sooner than he thought. Physicist LeóSzilárd is supposed to have read Wells's book and patented the idea. Szilárd was later directly involved in the Manhattan Project, which led to the tragedy of nuclear bombs being dropped on Japan in 1945.

James Gunn wrote in The Science of Science-Fiction Writing about Dr.H.G.Wells, —He believed that it was possible, through the use of what he first called "inductive history" and later "Human Ecology" (defined as the working out of "biological, intellectual, and economic consequences"), to chart the possibilities of the future and to push people into making sensible use of those possibilities. He was the first futurologist, the man who invented tomorrow,

In his _ "Invisible Man"' he predicts the possibility of humans having gained the technology of becoming invisible by using refractive index of light but it is not pure science fiction and makes sense scientifically. Technologists and scientists are working on materials today that have the property to bend light around an object making the object disappear or using lenses that curve light around the object making it disappear. Still in early days the invisibility now can only be possible at specific spots and looking through small apertures and specific angles.

The time machine, an invention introduced in a 1895 novella of H.G. Wells has yet to become a reality but theoretical physicists are working in this field too. There is a feeling that going back and forward in time can be made possible during experiments in particle physics or at the atomic levels but not possible at least till now at the macro level.

Known as the father of the modern submarine, American inventor Simon Lake had been captivated by the idea of undersea travel and exploration ever since he read Jules Verne' s Twenty Thousand Leagues Under the Sea in 1870.

Lake' s innovations included ballast tanks, divers' compartments, and the periscope. His company built the Argonaut—the first submarine to operate successfully in the open ocean, in 1898—earning him a congratulatory note from Verne.

While Jules Verne is perhaps most famous for his fictional submarine, the Nautilus, the French author also envisioned the future of flight.

Igor Sikorsky, inventor of the modern helicopter, was inspired by a Verne book, Clipper of the Clouds, (Robur the Conqueror) which he had read as a young boy. Sikorsky often quoted Jules Verne, saying —Anything that one man can imagine, another man can make real.

Robert H. Goddard, the American scientist who built the first liquid-fueled rocket—which he successfully launched on March 16, 1926—became fascinated with spaceflight after reading an 1898 newspaper serialization of H.G. Wells' s classic novel about a Martian invasion, War of the Worlds.

Goddard has said that concept of interplanetary flight —gripped my imagination tremendously. The innovators behind objects like the cellphone or the helicopter took inspiration from works like —Star Trek and War of the Worlds.

Director of research and development at Motorola is supposed to have credited the —Star Trek communicator as his inspiration for the design of the first mobile phone in the early 1970s.

"That was not fantasy to us", Martin Cooper is supposed to have said, —that was an objective.

French science fiction writer Jules Verne's who was one of the pioneers of science fiction writing imagines a lunar exploration mission more than a century before America sent manned spacecraft to the moon. Strangely there are several parallels drawn to the real life happening to what Verne wrote hundred years ago.

He wrote that United States would launch the first vehicle to go to the moon.

Verne's story has three astronauts like the actual Apollo 11. Not only that, the spacecraft of Jules Verne also resembles the future command modules.

Almost as if Verne witnessed Apollo moon landing his spacecraft uses retro-rocket to de accelerate the craft to descend slowly so that they do not crash onto the moon' s surface due to the gravitational pull.

Interestingly like in real life, Texas and Florida competed to host the launch site

A competition for the launch site would ensue between Florida and Texas which actually was resolved in Congress in the 1960s with KSC as the Flordia launch site and Houston, Texas as the Mission Control Center.

Verne predicted weightlessness and predicted that the astronauts on their return journey splash down same area of the Pacific Ocean.

All of this is described 106 years before the Apollo 11 mission.

Jules Verne in his book Twenty thousand Leagues Under the Sea published in 1879, created a huge electric submarine called Nautilus. It was a type of marine transport that wasn't invented until 1960, although there were prototypes that could have inspired Jules.

It was Jules Verne who in his work published in 1889 and writing of the Year 2889 describes the —phonotelephote, a forerunner to videoconferencing where one can see people and converse remotely. The device allowed —the transmission of images by means of sensitive mirrors connected by wires, writes Verne. This was one of the earliest references to a videophone in fiction.

Aldous Huxley's dystopian sci-fi novel 'Brave New World', set in a totalitarian future World State, anticipates and predicts advances in social conditioning and reproductive technology.

Isaac Asimov , End of Eternity (1955) is a time-travel story which has many concepts which we are only beginning to understand now and that too only in the world of quantum physics particularly parallel universes, violation of causality.

In 2001 : A Space Odyssey (1968), Arthur C Clarke predicted a number of technologies which were not in existence in the late 1960s.

Video calls and Skype video calls are taken for granted today, but it seemed like future fantasy when Stanley Kubrick and Arthur C. Clarke imagined it back in 1968 when it appeared in the seminal movie 2001: A Space Odyssey.

We're used to Siri-like interactive computers and the International Space Station now, but director Stanley Kubrick predicted them decades earlier.

His film 2001: A Space Odyssey is full of futuristic tech that would eventually exist, including iPad-like tablets (which also existed in a similar form PADD (Personal Access Display Device)on the original Star Trek.

The film made by Stanley Kubrick and Clarke in the book shows tablet computers, teleconferencing, robotic satellites, face and voice recognition, orbital space stations, and Artificial Intelligent robot like Hal.

Progress in the field of medicine and medical technology has been predicted by science fiction writers. The Tricorder of Star trek, the hand-held scanners used by crew members of the Starship for medical, security checks have only nowstarted being used like the Scanadu Scout a small disk that you can hold between your thumb and forefinger which embedded with scanners instantly begins registering physiological information as soon as you place the device to your forehead. Designed by Yves Behar, the lightweight, portable device represents a breakthrough in self-diagnosis, data mining and e-health technology.

In 2014, Organovo, a California biotech firm, took the capabilities of three-dimensional printing to an entirely new level by announcing its intention to begin selling 3D-printed liver tissue to allow for medical research on non-human organs. At the same time, the company's executive vice president predicted that in just a few years, such 3D tissue could be introduced into a human patient to encourage cell regeneration or replace small portions of existing tissue.

In Star Trek one is fascinated to see the Replicator, which using similar technology like the Transporter to dematerialize and then rematerialize matter in another form. The technology was used in the —Star Trekuniverse for everything from food to machinery to clothing.

One of the most ingenious, and little-recognized, science fiction films was —Gattaca, which when released in 1997 foretold of a future where children could be genetically manipulated to guarantee they received the best hereditary traits of their parents. The film is both thrilling in its depiction of future society and cautionary in its exploration of the genetic discrimination that might result when humans get the power to manipulate genes. As predicted the genetic manipulation of humans is slowly becoming an important issue today.

The Food and Drug Administration in 2014 approved a robotic device called ReWalk to assist individuals with lower body paralysis from a spinal cord injury. The motorized brace is fitted to support both legs and a portion of the upper body and uses motors to power movement at the hips, knees and ankles. ReWalk also includes a backpack with a computer and power supply. Use of the device would allow paralyzed individuals to sit, stand and walk with assistance.

These examples of technology merging with medical breakthroughs are just the beginning. Several predictions of Isaac Asimov's " I, Robot" over 60 years ago are coming true now as artificial intelligence and robots are slowly but surely becoming part of our life.

Robots as security guards, robots as industrial workers, robots as care givers, as cleaners in the house. A butler robot than can do simple tasks like wash dishes, fold clothes and vacuum. Robots that can keep track of vital signs of health and recognize warning signs.

According to Hiroshi Ishiguro, the Osaka University roboticist, robots will integrate into our lives the same way smartphones did a decade ago.

As human-robot interface increases and artificial intelligence of robots becomes highly sophisticated the boundaries between a live pet and artificial robot may start disappearing.

Many moral, social, emotional, and ethical questions to deal with robots may crop up as Asimov points out.

Philip K. Dick's 1958 sci-fi short story The Minority Report which shows the touch screen technology without the screen is slowly becoming a reality.

Interestingly Edward Bellamy's 1888 Utopia depicts the protagonist Julian West falling asleep in 1887 in Boston, Masschusetts and waking up in 2000 to find cards are used as money like 'debit'cards of today. "You observe, he pursued as I was curiously examining the piece of pasteboard he gave me, that this card is issued for a certain number of dollars. We have kept the old word, but not the substance…The value of what I procure on this card is checked off by the clerk, who pricks out of these tiers of squares the price of what I order."

Debit cards and credit cards would be invented more than 60 years later.

The credit card was invented in the year 1950, and is already being replaced by payment through mobile devices.

In the book 'Gulliver's Travels', published in 1735, Jonathan Swift predicted that Mars has two moons based on the astronomer Johannes Kepler's hypothesis in the early seventeenth century.

One hundred and fifty years later, Phobos and Deimos satellites were discovered, since the optics available back then didn't allow to see celestial bodies so small and so close to the planets.

Hugo Gernsback in his book 'Ralph 124C 41+', mentioned a device called —telephot, which allowed people to see each other while talking from long distances; a precursor to the current Skype.

Written in 1914, the novel also introduces the concept of radar, literally described as "a wave of pulsating polarised ether that is reflected in metallic objects and returns to the emitter, thus allowing to calculate the position and distance."

In his book '1984', George Orwell establishes a multitude of parallels between a fictional totalitarian and repressive society and that of today.

In the '2001: A Space Odyssey' movie, there is a scene in which astronauts watch/read from a pair of flat-screen tablets.

Not only is the tablet concept perfect, but the design is exactly the same as the ones we already use every day. Don't these devices look like iPads?

In the novel _2001: A SpaceOdyssey', Arthur C. Clarke discussed about a network of geo-synchronised satellites, which move around the Earth at the same speed, remaining in the same position and thus allowing global communication.

Clarke's exact description is closer to the technology of communications satellites, which were put into orbit for the first time 15 years after the publication of his book.

In 'Farenheit 451', Ray Bradbury describes a particular microphone headset that allows individuals to talk to each other way back in 1953.

Neither current telephone handsets nor Bluetooth communication was marketed until 2001.

In his first work, 'Neuromancer', released in 1984, William Gibson mentioned the term 'cyberspace' and predicted the Internet phenomenon and virtual reality.

Although the movie 'Total Recall' portrays a rather exaggerated view of the future, one of the technologies presented was very accurate.

The characters in the movie often use Johnny Cabs, taxis driven by automated drivers who control the vehicle as the movie predicted self-driving cars will be on the roads soon..

From the Internet to iPads to smart machines, some of the world's greatest advances in technology were once fictional speculation.

As sci-fi author Arthur C. Clarke wrote in Profiles of the Future (1962), 'The only way of discovering the limits of the possible is to venture a little way past them into the impossible.'